D1545409

Environmental Taxation and Climate Change

CRITICAL ISSUES IN ENVIRONMENTAL TAXATION

Series Editors: Larry Kreiser, *Cleveland State University, USA*, Julsuchada Sirisom, *Mahasarakham University, Thailand*, Hope Ashiabor, *Macquarie University, Australia* and Janet E. Milne, *Vermont Law School, USA*

The *Critical Issues in Environmental Taxation* series provides insights and analysis on environmental taxation issues on an international basis and explores detailed theories for achieving environmental goals through fiscal policy. Each book in the series contains pioneering and thought-provoking contributions by the world's leading environmental tax scholars who respond to the diverse challenges posed by environmental sustainability.

Previous volumes in the series:

Original book published by CCH Incorporated

Volumes I–IV published by Richmond Law Publishers

Volumes V–VIII published by Oxford University Press

Titles in the series published by Edward Elgar:

Volume IX   Environmental Taxation in China and Asia-Pacific
            Achieving Environmental Sustainability through Fiscal Policy
            *Edited by Larry Kreiser, Julsuchada Sirisom, Hope Ashiabor and
            Janet E. Milne*

Volume X   Environmental Taxation and Climate Change
            Achieving Environmental Sustainability through Fiscal Policy
            *Edited by Larry Kreiser, Julsuchada Sirisom, Hope Ashiabor and
            Janet E. Milne*

Environmental Taxation and Climate Change

Achieving Environmental Sustainability through Fiscal Policy

Edited by

Larry Kreiser

Professor Emeritus of Accounting, Cleveland State University, USA

Julsuchada Sirisom

Lecturer of Accountancy, Mahasarakham University, Thailand

Hope Ashiabor

Associate Professor of Law, Macquarie University, Australia

Janet E. Milne

Professor of Law, Vermont Law School, USA

CRITICAL ISSUES IN ENVIRONMENTAL TAXATION, VOLUME X

Edward Elgar

Cheltenham, UK • Northampton, MA, USA

© The Editors and Contributors Severally 2011

All rights reserved. No part of this publication may be reproduced, stored in a retrieval system or transmitted in any form or by any means, electronic, echanical or photocopying, recording, or otherwise without the prior permission of the publisher.

Published by
Edward Elgar Publishing Limited
The Lypiatts
15 Lansdown Road
Cheltenham
Glos GL50 2JA
UK

Edward Elgar Publishing, Inc.
William Pratt House
9 Dewey Court
Northampton
Massachusetts 01060
USA

K 3585.5

.E59

2011

o7332 30262

A catalogue record for this book is available from the British Library

Library of Congress Control Number: 2011926849

MIX
Paper from
responsible sources
FSC   FSC® C018575
www.fsc.org

ISBN 978 0 85793 786 5

Typeset by Columns Design XML Ltd, Reading
Printed and bound by MPG Books Group, UK

Contents

v

Figures

Tables

Editorial review board

The 15 chapters in this book have been brought to publication with the help of an editorial review board dedicated to peer review. The four members of the board are committed to the field of environmental taxation and are active participants in environmental taxation events around the world.

Lead Editor

Larry Kreiser
Cleveland State University, USA

Co-Editors

Julsuchada Sirisom
Mahasarakham University, Thailand

Hope Ashiabor
Macquarie University, Australia

Janet E. Milne
Vermont Law School, USA

Contributors

Batt, H. William, Robert Schalkenbach Foundation and International Union for Land Value Taxation, USA.

Black, Celeste M., University of Sydney, Australia.

Braathen, Nils Axel, Principal Administrator, Organisation for Economic Co-operation and Development (OECD), Environment Directorate, France.

Carraro, Fiorenza, University of Pavia, Italy.

Cela, Enian, Hiroshima University, Japan.

de Cendra de Larragán, Javier, University College London, UK.

Gee, David, European Environment Agency, Denmark.

Geekie, John T., Cleveland State University, USA.

Joseph, Sally-Ann, University of New South Wales, Australia.

Kaneko, Shinji, Hiroshima University, Japan.

Kennedy, Amanda, University of New England, Australia.

Kwon, Hai Sook, Kyung Hee University, USA.

Lee, Paul J., Cleveland State University, USA.

Panella, Giorgio, University of Pavia, Italy.

Park, Seung-Joon, Kyoto Sangyo University, Japan.

Phromlah, Wanida, University of New England, Australia.

Sirisom, Julsuchada, Mahasarakham University, Thailand.

Speck, Stefan, European Environment Agency, Denmark.

Sprohge, Hans, Wright State University, USA.

Tavallali, Rahmat O., Walsh University, USA.

Weber, Rolf H., University of Zurich, Switzerland.

Yamazaki, Masato, National Institute of Advanced Industrial Science and Technology (AIST), Japan.

Zatti, Andrea, University of Pavia, Italy.

Foreword

Many hundreds of environmentally related taxes are applied in OECD countries and elsewhere around the world (cf. www.oecd.org/env/policies/database), addressing a broad spectrum of environmental problems, such as waste generation, use of hazardous chemicals, traffic congestion, etc. – and, of particular relevance for this book, greenhouse gas emissions. There are many good reasons for their use, both from a theoretical and practical point of view.

One well-known argument for their use is that they (in principle) can equalise the marginal cost of reducing emissions across polluters, which, in the context of climate change mitigation, would result in a given level of carbon emissions being obtained at the lowest possible cost to society as a whole. Those who can reduce their emissions at a cost per unit lower than the tax rate set will do so – and those who face higher abatement costs will pay the tax instead of abating. Importantly, the taxes give the polluters a large flexibility to find low-cost ways of reducing their emissions. It would in practice be an impossible task for a government to achieve a similar outcome by telling each polluter how much – and how – they should abate.

In practice, however, governments rarely design environmental taxes in a way that provides an equal marginal abatement incentive to all sources contributing to a given problem – and this is also the case as regards carbon-related taxes. Certain polluters are often completely exempted, or face reduced tax rates; tax rates of 'carbon taxes' often vary between energy products in ways not reflecting their carbon content, etc. Two major reasons for this are fears of reduced international competitiveness of the most affected firms or sectors and fears that low-income households could be 'hit' particularly hard. While the merits of such concessions will always be arguable, there are several ways to address these problems while preserving the environmental integrity of the taxes applied. Providing compensation to the 'losers' via mechanisms outside of the tax itself is often preferable – as is broad international co-operation with respect to global problems, like climate change.

In addition to the *static efficiency* argument for the use of environmentally related taxes, such taxes also provide *dynamic efficiency*, giving polluters (and others) continued incentives for finding cheaper ways of reducing the emissions – e.g. via new inventions or via new uses of existing technologies. Such innovations provide polluters with a means to reduce their costs, causing total emissions to decrease. While this has been known in theory for a long time, a recent OECD book on *Taxation, Innovation and the Environment* (see www.oecd.org/env/taxes/innovation) provides much new empirical evidence in this regard.

From a more practical point of view, many environmentally related taxes are perhaps primarily applied in order to raise revenue – for a particular purpose or for public coffers in general. Examples include taxes applied to finance road building and maintenance (which can have rather mixed impacts on the environment in general and on CO_2 emissions in particular); to finance safe treatment systems for hazardous wastes; to raise revenues that allow a reduction in other taxes, e.g. on income; or, in the aftermath of the economic crisis that many countries have experienced in recent years, to limit public deficits. Regardless of the purposes to which such revenues are applied, the price impact of the taxes will always, in isolation, provide an incentive to reduce emissions.

Concerns exist that if environmental taxes are serving their intended purpose and reducing pollution, the revenues they yield will diminish over time. In the end, this is a question of the own-price elasticity of the tax-base in question. For some tax-bases, it is relatively easy to find substitutes – or to change behaviour in other ways – so a tax increase could rapidly cause an erosion of the tax-base. In the context of climate change mitigation, however, available estimates indicate that the long-term own-price elasticity of fossil fuel use is lower than one in absolute value (but certainly different from zero). Hence, a general increase in taxes on fossil fuels would cause the use of these fuels to be lower than otherwise – but the amount of revenues raised would increase.

We know a lot about the benefits of using environmentally related taxes and about how they work in practice. The chapters in this volume contribute further to this knowledge. In spite of a significant literature also on the 'political economy' of environmental policies in general, and of environmentally related taxes in particular, there is still much scope for further work on how best to help policy makers broaden the use of well-designed taxes (and emission trading systems) to address a wide range of environmental problems, climate change mitigation amongst them.

Nils Axel Braathen
Principal Administrator, Organisation for Economic Co-operation
and Development (OECD), Environment Directorate, France

Preface

Sometimes it is said that 'the best way to predict the future is to help shape it.'

We can help shape an environmentally sustainable future by designing more environmentally friendly taxation systems.

Volume X of Critical Issues in Environmental Taxation contains 15 chapters which provide insights and analysis for achieving environmental sustainability through fiscal policy. The main emphasis of Volume X is on environmental taxation and climate change.

The chapters are written by environmental taxation scholars from around the world. We hope you find their ideas to be interesting, thought-provoking, and worthy of serious consideration by policy makers.

Larry Kreiser, Lead Editor

Julsuchada Sirisom, Co-Editor

Hope Ashiabor, Co-Editor

Janet E. Milne, Co-Editor

September 2011

Abbreviations

AAU	Assigned Amount Unit (New Zealand)
AGS	Afforestation Grant Scheme (New Zealand)
AIDS	Almost Ideal Demand System
AIST	National Institute of Advanced Industrial Science and Technology (Japan)
AOC	Approximate original contour
ATO	Australian Taxation Office
BAU	Business as usual
BF–BOF	Blast furnace–basic oxygen furnace
BLDTF	Black Lung Disability Trust Fund (United States)
BTA	Border tax adjustment
CCS	Carbon capture and storage
CDM	Clean Development Mechanism
CDM EB	Clean Development Mechanism Executive Board
CER	Certified emission reduction
CES	Constant elasticity of substitution
CGE	Computable general equilibrium
CO_2	Carbon dioxide
CO_2-e	CO_2-equivalent
COMETR	Competitiveness Effects of Environmental Tax Reform
COP	Conference of the Parties
CPRS	Carbon Pollution Reduction Scheme (Australia)
DEQ	Department of Environmental Quality (United States)
DKr	Danish Kroner
D.P.R.	Italian Republic Presidential Decree
EAF	Electric arc furnace
EC	European Commission
EEA	European Environment Agency
EEB project	Energy efficiency in building project

EFR	Environmental fiscal reform
EPA	Environmental Protection Agency (United States)
ESCAP	Economic and Social Commission for Asia and Pacific (United Nations)
ETR	Environmental tax reform
ETS	Emissions Trading System
EU	European Union
FE model	Fixed Effect Model
GATS	General Agreement on Trade in Services (World Trade Organization)
GATT	General Agreement on Tariffs and Trade (World Trade Organization)
GDP	Gross domestic product
GFC	Green Fiscal Commission (United Kingdom)
GGAS	Greenhouse Gas Reduction Scheme (New South Wales, Australia)
GHG	Greenhouse gas
GLRC	Great Lakes Regional Collaboration (United States)
GRAM	Global Resource Accounting Model
ICI	Local council property tax (*Imposta Comunale sugli Immobili*) (Italy)
IEA	International Energy Agency
IMF	International Monetary Fund
IPCC	Intergovernmental Panel on Climate Change
ISPRA	Institute for Environmental Protection and Research (*Istituto Superiore per la Protezione e la Ricerca Ambientale*) (Italy)
ITAA	Income Tax Assessment Act (Australia)
MAF	Ministry of Agriculture and Forestry (New Zealand)
MIS	Managed investment scheme
MRV	Monitorable, reportable and verifiable
NAMA	Nationally appropriate mitigation action
NGAC	Greenhouse Abatement Certificate (New South Wales, Australia)
NSW	New South Wales (Australia)
NZ ETS	New Zealand Emissions Trading Scheme
NZU	New Zealand Unit

OECD	Organisation for Economic Co-operation and Development
OSM	Office of Surface Mining (United States)
PETRE	Resource Productivity, Environmental Tax Reform and Sustainable Growth in Europe
PFSI	Permanent Forest Sink Initiative (New Zealand)
ppp	Purchasing power parity
R&D	Research and development
RE model	Random Effect Model
REDD	Reducing Emissions from Deforestation and Degradation
SMCRA	Surface Mining Control and Reclamation Act (United States)
SNLT	Sector no-lose targets
SOP	Standard operation procedure
SSC	Social security contribution
SUR	Seemingly Unrelated Regressions
UGB	Urban growth boundary
UNEP	United Nations Environment Programme
UNFCCC	United Nations Framework Convention on Climate Change
VIF	Variance Inflation Factors
VOC	Volatile organic compound
WBCSD	World Business Council on Sustainable Development
WTO	World Trade Organization

PART I

Environmental Taxation Overview

1. Carbon-related taxation in OECD countries and interactions between policy instruments

Nils Axel Braathen[1]

INTRODUCTION

The use of *economic* instruments to address greenhouse gas emissions has widened in recent years, for example, in the form of 'carbon taxes', CO_2-related tax rate differentiation of motor vehicle taxes and emission trading systems. As climate change is a global problem, ideally all sources of emissions of greenhouse gases in general, and CO_2 in particular, ought to face a similar incentive at the margin to abate emissions.

This chapter discusses in Section I current use of carbon-related taxation in Organisation for Economic Co-operation and Development (OECD) countries, focusing in particular on the taxation of motor vehicle fuels and motor vehicles, and their impacts on the fuel-efficiency of the transport sector. In Section II, important interactions between a cap-and-trade system and other policy instruments meant to reduce CO_2 emissions are discussed.

I THE USE OF CARBON-RELATED TAXATION IN OECD COUNTRIES

I.1 Carbon-related Taxation of Energy Products

I.1.a Taxes on petrol and diesel

Most countries in the world levy excise taxes on motor vehicle fuels.[2] Figure 1.1 illustrates the total excise tax rates that applied to petrol and diesel in all OECD countries as of 1 January 2000 and 2010 respectively.[3] When these taxes were first introduced, few people worried about emissions of CO_2 or other greenhouse gases. Nevertheless, these taxes do have an impact on the prices of petrol and diesel – and hence on the use of these products. Indirectly they thus represent a sort of 'carbon tax'.

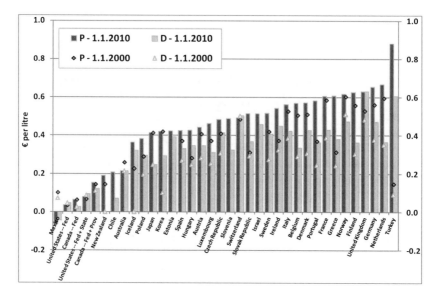

Notes: The graph includes all excise taxes levied on petrol and diesel, but not VAT or general sales taxes. 2009 average market exchange rates were used to calculate the tax rates in Euro for both years. P: Petrol; D: Diesel.
* In Mexico, the user price of motor vehicle fuels is kept almost stable over time, which in years with high crude oil prices has resulted in the excise tax on petrol and diesel turning into a subsidy.

Source: OECD/EEA database on instruments used for environmental policy, www.oecd.org/env/policies/database.

Figure 1.1 Taxes on petrol and diesel in OECD countries 01.01.2000 and 01.01.2010

However, the tax rates rarely reflect the carbon content of the two fuels. While the use of one litre of petrol causes 2.34 kg of CO_2 emissions, the use of one litre of diesel causes 2.68 kg of CO_2 emissions. In spite of this, in all OECD countries – except Switzerland, the United Kingdom and the United States – the tax rate per litre diesel is lower than the tax rate per litre petrol.[4]

By far the highest rate of tax on petrol in the OECD and the second highest rate on diesel are levied in Turkey – one of the OECD countries with the lowest income per capita.[5] Several motivations can explain this, but the environmental impacts are clear.

If the motor fuel taxes had been levied *only* to address greenhouse gas emissions, a tax rate of €0.6 per litre petrol would correspond to a 'carbon tax' of €256 per tonne of CO_2 emitted. A tax rate of €0.6 per litre diesel would similarly correspond to a 'carbon tax' of €224 per tonne of CO_2 emitted. However, in practice, the taxes on petrol and diesel are of course levied for a number of other reasons,[6] and one should not count all of them as 'carbon taxes'. Nevertheless, it is the full rate of tax that will influence the extent to which CO_2 will be emitted.

I.1.b 'Carbon taxes'

A handful of OECD countries have also implemented so-called 'carbon taxes' – but, while they can certainly represent a step in the right direction, these taxes do *not* normally impose a uniform level taxation of CO_2 emissions. Instead, the carbon taxes are designed as 'normal' taxes on (fossil) *energy products*,[7] with tax rates often varying across different energy products, and/or across different users of the energy products. It is also important to take into consideration any other taxes levied on the same energy products – and whether any changes were made to those taxes when the 'carbon tax' was introduced or modified.[8]

Table 1.1 compares the 'carbon tax' rates applied in six European countries and the total taxes levied on some selected energy products.[9] One can e.g. notice that even if Sweden has by far the highest nominal carbon tax rate, the total tax on natural gas is higher in Denmark than in Sweden (and in the other countries), and that the tax rate on coal in Denmark is only slightly lower than the (highest) one in Sweden.

Table 1.1 Carbon taxes and total taxes on selected energy products in OECD countries, 01.01.2010

	Denmark	Finland	Iceland	Ireland	Norway	Sweden	UK
Only 'carbon tax', per tonne CO_2	~€20	~€30–50	~€13	~€15	~€10–40	~€100	~€5–30
Heating oil, domestic use, per litre	0.33	0.087	0.02	0.04	0.17	0.41	0.0
Coal, per tonne	270.8	50.5	0.0	4.18	0.0	278.2	14.4
Natural gas, per Nm^3	0.35	0.02	0.0	0.03	0.01	0.24	0.02
Natural gas, per MWh	31.9	2.1	0.0	2.8	4.9	21.4	1.80
Petrol, per litre	0.57	0.63	0.36	0.54	0.62	0.52	0.63
Diesel, per litre	0.43	0.36	0.32	0.45	0.47	0.41	0.63

Note: The comparison should be used with caution. Whereas the first row only reflects the so-called 'carbon taxes', the rows below include *all excise taxes* levied on the energy products listed. Tax rates for natural gas are given both per standard m³ and per MWh, assuming that 1 m³ equals 0.01102 MWh.[10] The tax rates indicated for coal and natural gas in the UK do not apply to domestic use, which is fully exempted. For businesses that have entered so-called 'climate change agreements', an 80% tax reduction is available. No natural gas is used on Iceland, and heating oil is rarely used for domestic purposes. The tax rate for natural gas in Norway concerns natural gas used for heating purposes. Important tax reduction possibilities are also available for businesses in Denmark and Sweden, cf. www.oecd.org/env/policies/ database for further details.

By taking into account the carbon content of the different fuels, one can compare the total implicit tax rate per tonne of CO_2 emitted for each fuel. This is done in Figure 1.2.[11] It is clear that the tax rates vary widely between energy products. The differences in tax rates across product categories are, however, smaller in Sweden (with the highest 'carbon tax' rate) than in the other countries, with Denmark in the 'second place' in this 'ranking'.

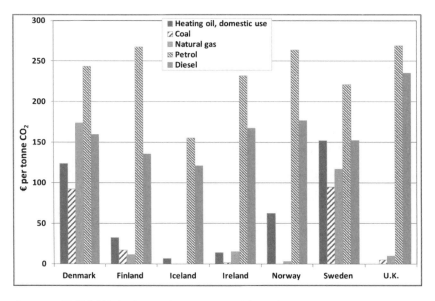

Source: OECD/EEA database on instruments used for environmental policy, www.oecd.org/env/policies/database, and author's own calculations.

Figure 1.2 Tax rates per tonne CO_2 emitted implicit in the excise tax rates on selected fossil fuels, 01.01.2010

I.2   Carbon-related Taxation of Motor Vehicles

A number of countries have also introduced carbon-related differentiation of the tax rates in their motor vehicle taxes; either taxes levied one-off in connection with the first purchase of the vehicle, or recurrent (annual) taxes that the owner of the vehicle has to pay in order to be allowed to use the vehicle. This section will give an overview of such taxes, building on OECD (2009a and 2009b).

I.2.a One-off motor vehicle taxes

Twelve OECD countries have one-off motor vehicle taxes with tax rates linked to the CO_2 emissions of each vehicle – or the fuel efficiency of each vehicle.[12] In Figure 1.3, the tax rate per (petrol-driven) vehicle is shown as a function of the amount of CO_2 the vehicle emits per km driven. It can be seen that there are *very* large variations in these taxes across countries – and between vehicles with different CO_2 emissions.[13] For vehicles with low emissions, some of the countries provide (sometimes large) subsidies, reducing the purchase price of the vehicles in question. For vehicles with CO_2 emissions exceeding 300 gram per km, the tax rates are *very* large in Norway, the Netherlands and Portugal.[14]

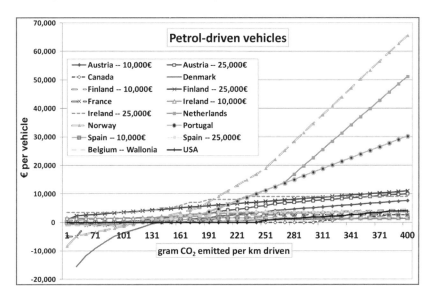

Figure 1.3 Tax per vehicle as a function of CO_2 emissions per km driven: petrol-driven vehicles, 01.01.2010

I.2.b Recurrent motor vehicle taxes

Eight OECD countries have carbon-related differentiation of recurrent motor vehicle taxes, and Figure 1.4 illustrates the amounts of taxes that are due per year, depending on the quantity of CO_2 emitted per km driven. Again, there are large differences across countries and according to CO_2

emissions. For vehicles emitting more than 250 gram CO_2 per km, the tax rate is particularly large for company-owned vehicles in France.

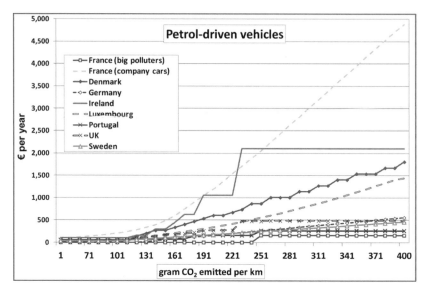

Figure 1.4 Tax per year as a function of the vehicles' CO_2 emissions per km driven: petrol-driven vehicles, 01.01.2010

By making assumptions regarding how many km a vehicle will be driven over its lifetime, and – in the context of the recurrent taxes – regarding the duration of the lifetime, one can calculate the tax rates per tonne of CO_2 the vehicle will emit over its lifetime. Assuming that each vehicle is driven 200 000 km in total, and that each vehicle has a lifetime of 15 years, Figure 1.5 illustrates the tax rates *per tonne of CO_2 over the vehicles' lifetimes* for selected emission-levels per km driven – adding together any one-off and recurrent taxes in each country.

It should in this context be emphasised that *each tonne* of CO_2 emitted causes the same environmental damage, regardless of whether the vehicle emits much or little. In other words, *a given tonne* emitted from a vehicle that causes high emissions per km driven does not harm the environment more than a tonne emitted from a vehicle causing low emissions per km. Nevertheless, Figure 1.5 illustrates that there are very large differences in the tax rates applied per tonne of CO_2 emitted between vehicles with different emission characteristics – and large differences between countries. The tax rates applied per tonne of CO_2 emitted from high-emitting vehicles are in some cases very large (both compared with the current price of an emissions allowance in the European Union's (EU) Emission Trading System (ETS) and the carbon prices estimated to be necessary for achieving relatively ambitions targets for CO_2 concentrations in the atmosphere)[15] – and it should of course be kept in mind that these CO_2 abatement incentives come on top of those provided through the motor fuels taxes.

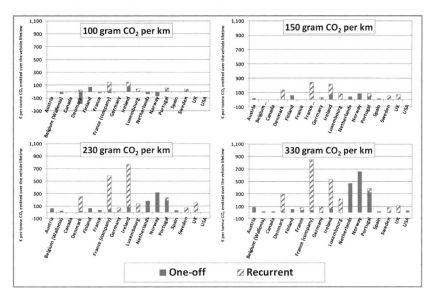

Figure 1.5 Tax per tonne CO_2 emitted over a vehicle's lifetime, for selected emission levels: petrol-driven vehicles, 01.01.2010

It is sometimes stated that reducing CO_2 emissions by making people buy low-emitting instead of high-emitting vehicles comes at no cost to society, because the two vehicles are such close substitutes. However, 'forcing' someone to buy a different vehicle type than what they would have bought in the absence of policy-interventions always cause a reduction in consumer surplus. One should consider carefully whether the benefits to society of doing so actually exceed this welfare loss.

II   INTERACTIONS BETWEEN A CAP-AND-TRADE SYSTEM AND OTHER POLICY INSTRUMENTS ADDRESSING CO_2 EMISSIONS[16]

The EU introduced its cap-and-trade based Emission Trading System (ETS) for CO_2 emissions in 2005, New Zealand enacted such a trading system in July 2010,[17] and similar trading systems have been under serious political discussion in a number of other countries. Such trading systems can achieve important emission reductions in a cost-effective manner, and an important body of literature discusses their application.[18]

However, parts of the literature overlook the interactions that can take place between a cap-and-trade system and other policy instruments.[19] When the cap is binding, this will determine the environmental outcome of the instrument mix directly – as long as it remains unchanged. Adding other policy instruments will not cause further emission reductions in the short term, but they *might* be justified through their ability to facilitate a lowering of overall emissions caps in the future.

II.1   Impacts in the Short Term, for a Given Cap

II.1.a   Impacts on CO_2 emissions

As mentioned, combining a CO_2-related cap-and-trade system with additional policy instruments that address emissions from the same sources as the trading system will not cause any extra emission reduction, as long as the cap of the trading system remains unchanged. The 'additional' instrument(s) would cause further abatement efforts by *some* emission sources, but – given the logic of a trading system – this would automatically lead to a reduction in permit prices and *increased emissions from some other source(s)* included in the trading system.

This point has *very* important implications for the environmental effectiveness and economic efficiency of many policy instruments applied in countries with a trading system in place, such as in all member countries of the EU. Given that the EU ETS, *inter alia*, covers electricity generation, additional policy instruments that address electricity use[20] or the CO_2 emissions caused by electricity generation[21] are unlikely to affect total CO_2 emissions on an EU-wide scale – in the short term.

For example, if country A spends €100 million to subsidise the installation of windmills to (partly) replace coal-fired power plants, this would cause a reduction in CO_2 emissions related to electricity generation in that country. This decrease in emissions would lower the demand of emission allowances in the EU ETS, causing allowance prices also to decrease – which, in turn, would make it cheaper for other firms covered by the system to emit CO_2. As long as the value of an emission allowance is positive, one or more of the other firms covered by the ETS will increase their emissions – up to the limits set by the total cap for the system. Hence, to achieve a reduction in CO_2 emissions at a EU (or global) level, it would be more efficient to spend the €100 million buying up emission allowances and cancelling them. That would guarantee a decrease in total emissions; the effect of which would be a decision by the country in question to reduce the total cap.[22]

Favourable feed-in tariffs and 'green certificates' are other policy measures EU countries use to promote the development of renewable sources electricity. Again, such measures will not have an impact on EU-wide CO_2 emissions, as long as the cap remains unchanged. In a discussion of a joint 'green certificates' scheme for Norway and Sweden, Bye and Hoel (2009) used the characteristic 'Green certificates – expensive and pointless renewable fun' (for policy makers).

The ban on incandescent light bulbs in EU countries has the same 'impact' on EU-wide CO_2 emissions in the short term.

Late in December 2009, the French Constitutional Court invalidated a 'carbon tax' that the Parliament had adopted a few weeks earlier – among other things, because the tax exempted firms covered by the EU ETS. Unfortunately, the Court had not understood that taxing those firms, while contributing to a reduction in *their* emissions, also would contribute to increasing CO_2 emissions elsewhere by a similar amount. Hence, a relatively well-designed 'carbon tax' may have been scrapped, at least partly owing to a misunderstanding of the economic effects of the measure on the part of the Constitutional Court.

The arguments above do *not* rest on an assumption that 'all markets always work perfectly'. There *are* certainly 'failures' in many of the markets covered by e.g. the EU ETS (imperfect competition, information failures, split incentives between landlords and tenants, etc.). But if the cap is binding, the workings of a trading system will 'automatically' lead to emission reductions in one place being countered by increased emissions from someone else covered by the system. It follows that supplementary policies measure designed to encourage emission reductions that are already covered by an emissions cap can only be justified to the extent that they address *other* market failures, cf. II.1.d below.

There is a 'reverse side to the coin' presented above: promoting the replacement of fuel-driven vehicles by electric vehicles *would* lead to a net reduction in EU-wide CO_2 emissions, regardless of the 'marginal' source of electricity within the EU. Because transport emissions are not presently covered by the ETS, some measures to encourage their reduction make sense. It is sometimes claimed that one needs to take into account the CO_2 emissions caused in the generation of the electricity used by the electrical vehicles, but – as long as the total cap of the ETS is unchanged – this is not correct. Such promotion would cause emission reductions equal to the emissions the fuel-driven vehicles would have caused – without any 'correction' related to the necessary electricity generation.[23,24]

II.1.b   Impacts on energy security

Even if the limited efficacy of adding other policy instruments to a cap-and-trade system is accepted, it is sometimes contended that such measures are desirable on the basis that they improve the energy security of a country or the wider region. For example, building wind turbines in EU countries is said to reduce the country's and the region's dependence on fossil fuel imports.

At first sight, this could seem correct. Wind turbines are driven by wind – not by fossil fuels. However, interactions with the 'cap-and-trade' system again come into play. As explained, replacing a coal- or gas-fired power plant with a windmill will necessarily *increase* CO_2 emissions from some other source(s) covered by the trading scheme. And these CO_2 *emission-increases* can only stem from increased *use of fossil fuels* among these 'other' sources. This could either be due to an increase in 'activity levels',[25] or to an increase in the CO_2 intensity of a given activity level[26] – or a combination of the two. Hence, overall fossil fuel use (measured in the amount of CO_2 emitted) in the EU as a whole would not be affected – as long as the cap is unchanged.

A caveat is that the 'supply risks' related to different fossil fuels could be (considered to be) different. If, for example, the supply of natural gas is considered more risky than the supply of coal, one might want to e.g. replace natural gas use in the electricity sector by windmills, and accept an increase in coal use in other electricity plants.

The EU has decided that 20% of total energy consumption is to come from renewable resources by 2020. However, due to the operation of the cap under the ETS this represents a moving target, which could prove to be very difficult and costly to meet. When *one* country takes action to comply with the obligation, the *same* country, or one of the *other* countries covered by the ETS, will automatically find it more difficult to reach the renewables target. This is because the fossil fuel replaced by a renewable energy source in one firm will largely be used somewhere else within the ETS. Hence, replacing e.g. a coal-fired power plant by a windmill park, or replacing fuel oil used in a paper mill by forest-based waste products, will free up CO_2 allowances, reducing their prices. The allowances will be used somewhere else – entailing increased fossil fuel use there. The total share of *renewables* in the energy mix – at the EU level – will thus not be *much* altered (and *total* energy use could increase). A given country may reduce its import dependence by such measures – but this would largely be at the 'expense' of some other country.

Hence, there are good environmental and economic reasons for EU authorities to review the relationship of the current renewables target to the CO_2 cap.

II.1.c Impacts on other types of emissions

Still another argument used to defend the use of 'flanking' instruments next to a cap-and-trade scheme for CO_2 emissions is that those instruments would *also* affect other types of emissions (e.g. SO_2, NO_x and particulates) stemming from the CO_2 emission sources. A first point to make is that it would most likely be more efficient to address such emissions through policy instruments specially designed for that purpose, than to rely on ancillary benefits from instruments designed to address CO_2 emissions.

More importantly, it is not given that there would be any net ancillary benefits. The reason is similar to the points made in the two preceding sub-sections: the 'flanking' instruments would cause increased use of fossil fuels 'elsewhere' among the sources covered by the trading system. The net impact on e.g. SO_2, NO_x and particle emissions will, hence, depend on the relative emission intensities of the sources that reduce and the sources that increase their CO_2 emissions.

II.1.d Impacts on economic efficiency

In spite of the arguments presented in the preceding sub-sections, there *can* be economic efficiency arguments for applying additional instruments next to a 'cap-and-trade' system if they effectively address relevant market failures, such as information barriers, market power in relevant markets, split incentives between landlords and tenants, etc.

However, these additional instruments should (like any other policy instrument) be subject to a careful cost-benefit analysis, where (almost) no benefits should be assumed to stem from reduced emissions of CO_2 or local air pollutants, or as regards an increase in energy security.

It is also well-known that there are positive externalities related to R&D and technological innovation. Potential innovators will not be able to capture all the benefits of innovation for themselves, and in an unregulated economy, too few resources would be devoted to such activities. There are, hence, economic efficiency arguments for applying instruments to promote new technologies that help abate CO_2 emissions, next to a cap-and-trade system – also while the cap remains unchanged – if the expected benefits exceed the expected costs.

II.2 Impacts on Future Caps, in the Medium to Long Term

Having so far discussed the impacts of 'flanking' instruments next to *a given cap* on CO_2 emissions, it is also important to address the impacts of these instruments on the level of 'strictness' of the cap in the future. Additional instruments certainly *can* create the conditions to tighten the cap in the future, but whether this actually occurs will depend on a range of factors, including the political economy conditions surrounding energy policy in the medium to longer term.

I will first focus on instruments meant to stimulate technological development, while also having an impact on current emission levels.[27] Examples include feed-in tariffs for renewables, green certificates and emission standards for electrical appliances. Part of their rationale in conjunction with cap-and-trade schemes is that they could make it possible to set a stricter cap in the future. As explained, such instruments will reduce emissions from some of the capped sources, free up emission allowances, causing allowance prices to decrease. The reduced allowance prices will (marginally) reduce the *incentives* for all other sources to develop new abatement technologies – the profitability of spending time and resources to develop such technologies will (marginally) be reduced.

The increase in emissions elsewhere will also mean that fewer abatement measures may be applied there in the short to medium term. Hence, among these other emission sources, there may be less 'learning by doing', marginally reducing their *ability* to innovate.

However, returning to 'flanking instruments' more broadly – it is important how they can be expected to affect current and future allowance prices.[28] The caps under the EU's 20-20-20 decision[29] were the result of a political process, rather than being necessarily based on careful estimation of the related costs. It is quite possible that future caps at least partly will be based on assessments of the expected costs of achieving them.

If an additional policy instrument effectively contributes to lowering current and/or expected future allowance prices, by helping to overcome current market failures in an efficient way, and/or by cost-effectively stimulating development of new abatement technologies, it *could* contribute to the setting of a stricter cap for future years. If not, it would pull in the opposite direction.

From a political economy perspective, policy makers have not felt it feasible to tighten the cap of the EU ETS enough to raise the price of carbon emissions to the level that would cause technological change on the scale needed to address the threat of major climate changes. In such a situation, it can make sense to try to facilitate the setting of a stricter cap in the future by applying 'flanking instruments' next to the ETS – but it is important to assess both *ex ante* and *ex post* whether the additional instruments in practice can be expected to lower the cost of reducing the capped emissions (or have done so).

II.3 Conclusion on Interactions Between Cap-and-Trade Systems and Other Instruments

To conclude this discussion: emission trading systems are (potentially) environmentally effective and economically efficient[30] instruments to address emissions of CO_2 and other greenhouse gases. Once a 'cap-and-trade' system has been put in place, further emission reductions are unlikely to be obtained by applying additional policy instruments to the same emission sources, as long as the cap is unchanged. If an additional instrument *in practice* contributes to reducing the costs of complying with the cap, it *could* contribute to a stricter cap being set in the future – to the extent that such considerations are taken into account when future 'caps' are set.[31]

Policy makers in countries with a 'cap-and-trade' system in place should consider carefully the actual contributions of any other policy instrument(s) they apply to address emissions from the same sources. There is a real danger that some of them will increase the total cost of reaching a given environmental outcome without making future reductions in the cap likely.

NOTES

[1] I thank several colleagues for helpful comments. The views expressed are, however, my own, and do not necessarily reflect the views of OECD or its member countries.

[2] New Zealand does not tax diesel fuel, but apply a tax per km driven by diesel-driven vehicles instead.

[3] In *real* terms, the (weighted) average tax rate on petrol in OECD member countries declined 8.1% between 2000 and 2010, cf. OECD (2010).

[4] On top of that, several other environmental damages are larger per litre diesel than per litre petrol, e.g. emissions of NO_x and particulate matter.

[5] Using a common exchange rate in 2000 and 2010, while the market exchange rate depreciated significantly, tends to over-emphasise the strong growth in the tax rates that in any case took place.

[6] For example, West and Williams (2007) found that motor fuels are complements in consumption to leisure. Taxing motor fuels hence make it possible to *indirectly* tax leisure, thus correcting a distortion otherwise difficult to address regarding the choice between work and leisure.

[7] As long as carbon capture and storage is not commercially available, there is generally a one-to-one relationship between CO_2 emissions and the carbon content of each fossil fuel. Differentiating the tax rates according to the carbon content of each fuel would hence be a close proxy for taxing the CO_2 emissions.

[8] For example, 'energy taxes' previously levied on a number of energy products were reduced when the 'carbon tax' in Sweden was increased in the early 1990s.

[9] The comparison does *not* take into account a number of exemptions, refund mechanisms and tax ceilings that apply in certain cases. Hence, it should be used with great caution.

[10] See www.carbontrust.co.uk/cut-carbon-reduce-costs/calculate/carbon-footprinting/pages/conversion-factors.aspx.

11. The carbon content in different qualities of coal can vary greatly. Here it is assumed that one kg coal causes emissions of 2.93 kg of CO_2, cf. http://en.wikipedia.org/wiki/Pit_coal. It is also assumed that one MWh of natural gas causes emissions of 0.184 tonne of CO_2, cf. www.carbontrust.co.uk/cut-carbon-reduce-costs/calculate/carbon-footprinting/pages/conversion-factors.aspx, and that heating oil causes the same CO_2 emissions as diesel.

12. Given the fuel used to drive the vehicle, there is a direct link between fuel efficiency and the CO_2 emissions. The CO_2 emissions and/or the fuel efficiency of the vehicles are estimated in specially designed test cycles but there is often a significant difference between these estimates and actual on-road emissions or fuel use.

13. In Austria, Finland, Ireland and Spain, the tax rate depends on the price of the vehicle. The graph shows – as *examples* – the tax rates for vehicles with pre-tax prices of €10 000 and €25 000 respectively. There are also differences in the taxes on petrol-and diesel-driven vehicles in some countries. The tax rates shown are for petrol-driven vehicles.

14. The graph only illustrates the CO_2 element of the vehicle taxes. In Norway, motor vehicles are *also* (heavily) taxed according their weight and motor power.

15. Cf. OECD (2009c).

16. Other papers discussing this issue include Sijm (2005), Sijm and van Dril (2003), Oikonomou and Jempa (2008), Pethig and Wittlich (2009), Böhringer and Rosendahl (2010), Fischer and Preonas (2010) and OECD (2011).

17. See www.climatechange.govt.nz/emissions-trading-scheme/about/.

18. See e.g. Weitzman (1974), Roberts and Spence (1976), Tietenberg (2006), OECD (2008) and Kaplow (2010).

19. For further discussion of instrument mixes for environmental policy, see OECD (2007) and Braathen (2007).

20. For example, measures to increase the energy efficiency of electrical appliances and taxes on electricity use.

21. For example, subsidies to promote renewable energy sources, and standards for the renewables content in electricity generation.

22. However, by contributing to higher permit prices, this *could* have a negative impact on future 'caps'.

23. There are hence arguments for putting in place *some* policy instruments aiming to promote the *replacement* of petrol- and diesel-driven vehicles by electric vehicles. However, the measures applied for this purpose in some countries seem out of proportion to the benefits achieved. For example, ECON (2009) indicates that subsidies given to electrical vehicles in Norway exceed €2500 per tonne CO_2 abated.

24. For individuals in the EU that would like to 'do something for the climate', replacing a fuel-driven car by an electric one could, hence, be an option. If the old car emitted 180 gram CO_2 per km, and it would be driven 200 000 km over its lifetime, 36 tonnes CO_2 would thus be avoided. But, if the price of an emission allowance in the EU ETS is €15, a similar environmental impact could be obtained for €540, by buying and cancelling 36 allowances.

25. For example, more fossil fuel-based power plants.

26. For example, substitution of a coal-fired power plant for a gas-fired one.

27. The points made here are *not* relevant for instruments that can trigger technology development, *without* having a (significant) impact on current CO_2 emissions, such as e.g. public subsidies for research in break-through technologies. If an instrument does not free up allowances in the short term, the 'perverse' counter-effects mentioned here will not materialise.

28. As Roberts and Spence (1976) pointed out, there *are* arguments for combining a trading system with a set of 'taxes'. If allowance prices were 'very low', a 'tax' could act as a floor on carbon abatements incentives – ensuring a continued incentive to innovate. If allowance prices were 'very high', emitters could have the option of instead paying a predetermined tax that would act as a ceiling on the allowance prices. This 'tax' would, however, not be paid for emissions for which the emitter holds an

allowance. However, limiting the allowance prices over a long period of time by putting in place a price ceiling would make it more difficult to achieve the given emission reductions.

[29.] In March 2007, the EU Heads of State and Government agreed a series of climate and energy targets to be met by 2020, known as the '20-20-20' targets. These are:
- a reduction in EU greenhouse gas emissions of at least 20% below 1990 levels;
- 20% of EU energy consumption to come from renewable resources;
- a 20% reduction in primary energy use compared with projected levels, to be achieved by improving energy efficiency.

See http://ec.europa.eu/clima/policies/package/index_en.htm for further information.

[30.] This is particularly the case if the permits are auctioned, and not handed out for free, cf. OECD (2008).

[31.] There is certainly scope for applying other instruments to address the sources not already covered by a trading system.

REFERENCES

Böhringer, Christoph and Knut Einar Rosendahl (2010), 'Green Promotes the Dirtiest: On the Interaction between Black and Green Quotas in Energy Markets', *Journal of Regulatory Economics*, Vol. 37, pp. 316–25.

Braathen, Nils Axel (2007), 'Instrument Mixes for Environmental Policy: How Many Stones Should be Used to Kill a Bird?', *International Review of Environmental and Resource Economics*, Vol. 1, pp. 185–235.

Bye, Torstein and Michael Hoel (2009), 'Green Certificates – Expensive and Pointless Renewable Fun', English translation of an article in *Samfunnsøkonomen* nr. 7, 2009. Available at www.ssb.no/english/research/articles/2009/12/1259932098.4.html.

ECON (2009), *Virkemidler for introduksjon av el- og hybridbiler* (Policy Measures for Introduction of Electrical and Hybrid Vehicles) [in Norwegian]. Report prepared for the Norwegian Petroleum Institute, Econ Pöyry, Oslo. Available at http://np.nsp01cp.nhosp.no/getfile.php/Filer/Tema/Miljo/Virkemidler%20for%20introduksjon%20av%20el-%20og%20hybridbiler%20okt09.pdf.

Fischer, Carolyn and Louis Preonas (2010), 'Combining Policies for Renewable Energy: Is the Whole Less Than the Sum of Its Parts?', *International Review of Environmental and Resource Economics*, Vol. 4, pp. 51–92.

Kaplow, Louis (2010), *Taxes, Permits and Climate Change*, Discussion Paper No. 675, Harvard Law School, available at www.law.harvard.edu/programs/olin_center/papers/pdf/Kaplow_675.pdf.

OECD (2007), *Instrument Mixes for Environmental Policy*, OECD, Paris. See www.oecd.org/env/policies/mixes.

OECD (2008), *Environmentally Related Taxes and Tradable Permit Systems in Practice,* OECD, Paris. Available at www.oecd.org/officialdocuments/displaydocumentpdf?cote=COM/ENV/EPOC/CTPA/CFA(2007)31/FINAL&doclanguage=en.

OECD (2009a), *The Scope for CO_2-based Differentiation in Motor Vehicle Taxes – In Equilibrium and in the Context of the Current Global Recession*, OECD, Paris. Available at www.olis.oecd.org/olis/2009doc.nsf/LinkTo/NT000068BA/$FILE/JT03271502.PDF.

OECD (2009b), *Incentives for CO$_2$ Emission Reductions in Current Motor Vehicle Taxes*, OECD, Paris. Available at www.olis.oecd.org/olis/2009doc.nsf/LinkTo/NT00004D96/$FILE/JT03269005.PDF.

OECD (2009c), *The Economics of Climate Change Mitigation: Policies and Options for Global Action Beyond 2012*, OECD, Paris. Available at http://dx.doi.org/10.1787/9789264073616-en.

OECD (2010), *Taxation, Innovation and the Environment*, OECD, Paris. See www.oecd.org/env/taxes/innovation.

OECD (2011), *Interactions between Emission Trading Systems and other Overlapping Policy Instruments*, OECD, Paris. Available at www.oecd.org/dataoecd/11/51/48188899.pdf.

Oikonomou, V. and C.J. Jepma (2008), 'A Framework on Interactions of Climate and Energy Policy Instruments', *Mitigation and Adaptation Strategies for Global Change,* Vol. 13(2), pp. 131–56.

Pethig, Rüdiger and Christian Wittlich (2009), 'Interaction of Carbon Reduction and Green Energy Promotion in a Small Fossil-Fuel Importing Economy'. *CESifo Working Paper Series*, No. 2749. CESifo Group, Munich. Available at www.ifo.de/pls/guestci/download/CESifo%20Working%20Papers%202009/CESifo%20Working%20Papers%20August%202009/cesifo1_wp2749.pdf.

Roberts, Marc J. and Michael Spence (1976), 'Effluent Charges and Licenses under Uncertainty.' *Journal of Public Economics,* Vol. 5, pp. 193–208.

Sijm, Jos (2005), 'The Interaction between the EU Emissions Trading Scheme and National Energy Policies', *Climate Policy*, Vol. 5, pp. 79–96.

Sijm, Jos and A.W.N van Dril (2003), *The Interaction between the EU Emissions Trading Scheme and Energy Policy Instruments in the Netherlands*. Report prepared for the European Commission. Available at www.ecn.nl/docs/library/report/2003/c03060.pdf.

Tietenberg, Tom H. (2006), *Emission Trading: Principles and Practice*. Second edition. Resources for the Future, Washington D.C.

Weitzman, Martin L. (1974) 'Prices vs. Quantities', *Review of Economic Studies*, Vol. 41(4), pp. 477–91.

West, Sarah E. and Roberton C. Williams III (2007), 'Optimal Taxation and Cross-price Effects on Labor Supply: Estimates of the Optimal Gas Tax', *Journal of Public Economics*, Vol. 91, pp. 593–617.

2. Implications of environmental tax reforms: revisited

Stefan Speck and David Gee

1. INTRODUCTION

During the last two decades several European countries implemented an environmental tax reform (ETR) (Speck and Jilkova, 2009). ETR is a public policy tool that applies revenue-raising economic instruments (which may be taxes or auctioned permits in an emissions trading scheme) to resource use (including energy) and pollution, in order to increase resource productivity, employment and innovation and to help improve the environment. The generated revenues can be used for different policy purposes. ETRs implemented in Europe are characterised as tax-shifting programmes and thereby closely following the revenue-neutrality principle as the additional generated revenues have been generally recycled back to the economy by reducing other more economically damaging taxes, such as labour or capital taxes.

ETR can be part of the public policy packages in developed countries, as well as in economies in transition and developing countries.[1] However, whilst much can be learned from country experiences with ETR, the lessons learned in individual countries cannot be directly transferred to other countries: ETR must be specifically designed for the fiscal, economic, social and environmental conditions of each country.

Experiences gained in several EU member states over 20 years show broadly positive results (EEA, 2005, and Barker *et al.,* 2009a). There has been little progress with ETR overall. In fact, at EU 25 level the share of environmental taxes has been decreasing and is now (2008) at the lowest level compared with 1995 (Eurostat, 2010). The main barriers to ETR are the potential loss of competitiveness when an ETR is implemented unilaterally (Ekins and Speck, 1999 and 2008, and Andersen and Ekins, 2009) and the equity issues that arise when environmental taxes fall disproportionately on low-income and rural households.

These topics are the main focus of this chapter. Several large research projects have been recently commissioned to analyse these aspects in more detail. The chapter summarises the key results of these projects in the context of current developments, such as the green economy, budget deficits and ageing, unequal societies.

2. THE DEFINITION OF ETR AND EFR

Environmental tax reform (ETR) – also sometime called green tax reform or ecological tax reform – has been implemented in several European Union (EU) member states, including Sweden, Denmark, Germany, the UK, Estonia and the Czech Republic. The concept of ETR is also actively promoted by international organisations, such as OECD (2005, 2010a and 2010b), World Bank (2005), United Nations Economic and Social Commission for Asia and Pacific (ESCAP, 2008), the EC (EC, 1993), and the European Environment Agency (EEA) (1996, 2000, 2005 and 2010). The underlying principle becomes clearer in the definition used in a report published by the EEA:

> ETR is a reform of the national tax system where there is a shift of the burden of taxation from conventional taxes, for example, on labour, to environmentally damaging activities, such as resource use or pollution. The burden of taxes should fall more on 'bads' than 'goods' so that appropriate signals are given to consumers and producers and the tax burdens across the economy are better distributed from a sustainable development perspective (EEA, 2005, p. 84).

The concept of an environmental fiscal reform (EFR) is broader as defined by reports published by the World Bank (2005) and the OECD (2005) as it 'refers to a range of taxation and pricing measures which can raise fiscal revenues while furthering environmental goals' (OECD, 2005, p. 24).

3. AN OVERVIEW OF THE KEY FINDINGS OF RECENT PROJECTS ASSESSING ETR IN EUROPE

3.1 Competitiveness Effects of Environmental Tax Reform (COMETR)

This research project was funded by the European Commission[2] and was implemented between 2004 and 2007. The project analysed the competitiveness impacts of the ETRs implemented in six EU member states

(Denmark, Finland, Germany, the Netherlands, Sweden and the UK) at a sectoral level as well as at the macro level by applying modelling frameworks (bottom-up and macro-economic). The overall perspective was ex-post but the modelling framework also included an ex-ante assessment. Evidence showed that the 'double dividend' theory proved true: five EU member states experienced reduced GHG emissions and increased employment via the introduction of CO_2/energy taxes and a reduction in the tax burden on labour[3]. In Sweden, Denmark, the Netherlands, Finland and Germany CO_2 and energy taxation over the last two decades has also made a small but positive contribution to economic growth of up to 0.5%, while CO_2 emissions have been reduced. In the UK the reform has minimal impacts, which can be attributed to the fact that the scale of the tax rates (i.e. climate change levy) levied and the revenues generated and recycled back has been rather modest, i.e. amounting to around 0.06% of gross domestic product (GDP) in 2004.

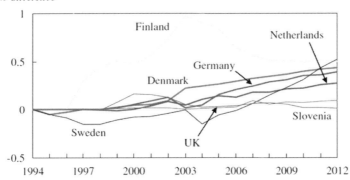

Note: % difference is the difference between the base case and the counterfactual reference case.

Source: Barker *et al.*, 2009a.

Figure 2.1 The effect of ETR on GDP

The positive contribution to economic growth arises because carbon-energy taxation leads to more efficient use of energy while at the same time the revenues from the energy-carbon taxes were recycled back to the economy by reducing either income taxes or social security contributions.

% difference

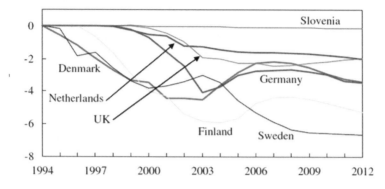

Note: % difference is the difference between the base case and the counterfactual reference case.

Source: Barker *et al.*, 2009a.

Figure 2.2 The effect of ETR on GHG emissions

3.2 Resource Productivity, Environmental Tax Reform and Sustainable Growth in Europe (PETRE)

The PETRE project[4] assessed the economic, environmental and resource implications, for Europe and the rest of the world, of a large-scale ETR which aimed to achieve the EU's greenhouse gas (GHG) reduction targets by 2020, i.e. an ex-ante assessment indicating the potential of an ETR to achieve a clear policy goal.

The project applied a range of methodologies, including two macro-econometric energy-environment-economy (E3) models[5] and physical input-output modelling, and the Global Resource Accounting Model (GRAM)[6] was constructed illustrating the European trade patterns from the perspective of material flows. The modelling exercise was also expanded by including the results of the material flows and this allowed studying how the introduction of an ETR in Europe would affect the economies and environments in other world regions.

The key environmental tax assumptions were:

- A carbon tax was introduced on all non-EU ETS sectors. The carbon tax rate was equal to the carbon price modelled in the EU ETS.
- The taxation of material inputs (biomass and minerals) apart from

energy products; materials taxes were are additional to existing taxes and were introduced at 5% of total price in 2010, increasing to 15% by 2020.

The ETRs were modelled in a revenue-neutral fashion as all revenues were recycled back to the economy: the revenues paid by industries were used for a reduction in employers' social security contributions (SSC) and the revenues paid by households financed an income tax reduction.

3.2.1 Economic results

It was estimated that the carbon prices needed to reach the EU GHG targets should be between 53 and 68 €/ton CO_2 depending on the scenario assumptions and model.[7]

Table 2.1 Aggregate impacts on GDP and employment (% difference from base in 2020)

Variable	Scenario			
	HS1	HS2	HS1	HS2
	GINFORS		E3ME	
Carbon price (€ per ton CO_2 in 2008 prices)	68	61	59	53
GDP	-0.6	-0.3	0.2	0.8
Employment	0.4	0.4	1.1	1.1
Consumption	-0.7	-0.5	0.6	0.7
Investment	-0.3	0.8	-0.3	0.3
Exports	-1.4	-1.3	-0.2	0.8
Imports	-0.8	-0.4	-0.2	0.2
CPI	0.9	1.1	0.8	0.7

Note: Figures show percentage change in EU 27 main indicators in 2020 compared with the relevant baseline for each scenario.

Source: Barker *et al.*, 2011.

The carbon prices in both models are in the same range and relatively high in particular (see Table 2.1) when compared with the prices seen so far in the EU ETS, which have been around €10 to €20 per ton of CO_2. The modelled carbon price does not take into account any supportive regulatory changes (e.g. in order to reach the EU 20% energy efficiency target and the EU's 20% renewables target in 2020). Furthermore the reduction target is met by emission reductions in just the EU.

A lower carbon price can achieve the same GHG reduction target if a part of the tax revenues (i.e. 10%) are invested in low carbon technologies. The effect on GDP was estimated to be small and positive in the E3ME modelling exercise and slightly negative by GINFORS. Both models showed that ETR would lead to a small increase in employment.

Overall, the results indicated that a broadly based ETR across Europe could play a very important and cost-effective role in meeting the EU's emission reduction targets for 2020. However, the impact of ETR would vary across EU member states, with both positive and negative impacts on different countries and sectors. Energy- and material-intensive sectors as well as the sectors intensively using these products tend to lose out from the ETR. In contrast, the labour-intensive sectors gained as they benefited from lower employment costs through the reduction in employers' SSC. In addition, sectors which are producing consumer goods are also gaining but to a lesser extent as they are benefitting from lower income tax rates. These findings are more or less consistent with the literature assessing the ETR (for examples see Barker *et al.*, 2009a, and Lutz and Meyer, 2009).

3.2.2   Revenue recycling and revenue stability

The ETR studied in this project raised large amounts of revenues as the additional revenues are in the same range or even exceed the 2005 level of environmental taxes (see the last column in Table 2.2 below). It is interesting to note that a rather large part of the additional tax revenues come from the materials tax and that the share of materials tax revenues in total ETR revenues could easily reach and exceed 40% in some countries (e.g. Denmark, Sweden and Spain). This finding is important as the revenues generated from taxes levied on energy products and from carbon pricing – including the auctioning of EU emission allowances – may decline in the long term as carbon emissions are reduced by climate policy, unless the tax rates is increased steadily to reflect increasing knowledge about the externalities of fossil fuels.

This exercise clearly demonstrated that the implementation of any real-life ETR has to carefully consider the specific contexts and situations of the countries in question and confirms the fact that there is no 'one fits all' ETR design, even if the general principles are relevant to all countries.

Table 2.2 shows how the recycling of the ETR revenues affecting the tax-to-GDP ratios. The analysis assumes that the overall fiscal structure in 2020 is the same as in 2005 and that the overall total tax to GDP ratio constant (column 2). The recycling of the ETR revenues lead to a reduction in both the personal income tax-to-GDP ratio and the SSC-to-GDP ratio. The higher reduction with regard to SSC is not surprising since a larger part of the ETR's revenues are paid by industry (including nearly all the materials taxes) and recycled back via the reduction of employers' SSC.

Table 2.2 Effects of the ETR on various tax-to-GDP ratios of selected EU member states

	Total tax incl. SSC	Personal income tax	SSC (paid by employer)	SSC (total)	Environ-mental taxes
	in % of GDP	in % of GDP	in % of GDP	in % of GDP	in % of GDP
	Fiscal structure of 2020 based on the fiscal structure of 2005 (as shown in brackets) and ETR				
Czech Republic	37.1	4.3 (4.6)	7.1 (10.3)	12.9 (16.1)	6.2 (2.7)
Estonia	30.9	5.5 (5.6)	8.6 (9.9)	9.0 (10.3)	3.7 (2.3)
Germany	38.8	8.2 (8.6)	5.1 (7.0)	14.4 (16.3)	4.8(2.5)
Portugal	35.1	4.9 (5.3)	5.7 (7.3)	9.8 (11.4)	5.1 (3.1)
Romania	27.9	2.1 (2.3)	5.3 (6.4)	8.6 (9.7)	3.3 (2.0)
Sweden	49.6	15.3 (15.5)	8.5 (9.8)	11.5 (12.8)	4.3 (2.8)
United Kingdom	36.1	9.7 (10.2)	2.2 (3.7)	5.2 (6.7)	4.5 (2.5)
EU 27	39.2	7.3 (7.7)	4.9 (6.7)	9.2 (11.0)	5.0 (2.8)

Note: SSC – social security contributions.

Source: Barker *et al.*, 2011 and Eurostat, 2010.

As mentioned above, one of the topics discussed in the context of an ETR is the question of the stability of environmental tax revenues. The principal arguments for introducing environmental taxes are to raise revenues more efficiently and to influence the behaviour of economic actors. If an environmental tax is being effective and achieving the objective of changing behaviour, then the tax revenues can fall. For example, the Irish plastic bag tax led to a dramatic fall in the use of plastic bags and associated revenues.

However, other environmental taxes, in particular those levied on energy products as well as transport taxes, have been reliable sources of substantial revenues for decades, which is essential for a successful tax-shifting ETR package. Furthermore, the effect on the revenues will depend on the proportion of the tax in the price of the affected products and the price elasticity of demand. If the price elasticity of demand is between 0 and -1, as mostly seems to be the case for energy products, then an increase in the tax on the products will still yield an increase in revenue. In addition, it is crucial to keep in mind that economic growth will tend to increase the demand for energy. This is of course one reason why an escalating tax is so necessary if energy demand and associated emissions are to be substantially reduced. Without a continually increasing energy price via taxes, growing incomes and the rebound effect from energy efficiency improvements are likely to result in increasing energy use in the future, as they have in the past (Sorrell, 2007, Barker *et al.*, 2009b and Ekins and Speck, 2011). Broadening the tax base with a materials tax, as it was done in the ETR scenario of the PETRE project, can also help to maintain revenues from environmental taxes. The results may be considered as quite robust as the model specifications and their parameterisation are based on empirical estimations.

3.3   The EEA Project on ETR

The European Environment Agency (EEA) funded a project which was based on the PETRE project analysing aspects of eco-innovation and ETR (Gehr *et al.*, 2009) and the equity effects of an ETR (Blobel *et al.*, 2009). In addition, a qualitative analysis was undertaken aiming to understand the scope for using ETR to address current and future environmental challenges as well as assessing possible paths to develop a more harmonised approach to ETR in the EU (Bassi *et al.*, 2009).

3.3.1   ETR, innovation and employment

The project aimed to get an understanding of what drives innovation and innovative behaviour in the economy, in particular by studying the literature on price instruments, especially environmental tax reform, and its effects on innovation (Gehr *et al.*, 2009). In addition, based on the modelling results of the PETRE result, scenarios were modelled analysing how ETR revenues can be used to foster innovation.

The literature review suggested:

> … that environmental regulation in general, and price-based policy instruments such as environmental taxes and investment subsidies in particular, can (in

theory) and do (in practice) have a positive impact on both innovation and diffusion of environmental technologies. However, the supporting empirical and case study evidence is not universal and the effectiveness of these instruments would appear to vary across different sectors and different types of innovation. (Gehr et al., 2009, p.23)

This result is more or less in line with the more recent findings of the OECD, concluding that 'Environmentally related taxation stimulates the development and diffusion of new technologies and practices' (OECD, 2010a, p.12). As with the EEA project, 'the case studies undertaken as part of this project do not provide unambiguous evidence that environmentally related taxation will always lead to innovation and the adoption of new technologies and processes' (OECD, 2010a, p.13). However, it should be noted that very few evaluations of complex policy interventions can provide unambiguous evidence.

The scenarios undertaken as part of the EEA project were built on the PETRE scenarios and aimed to study the impacts of eco-innovation and the possible role of international trade as well as analysing the effects of changing the input structure of the utility sector, i.e. from conventional electricity production to renewables. These two scenarios are closely linked to scenario HS2 of the PETRE project as 10% of the revenues are spent on eco-innovation measures. The results demonstrate that targeted spending on eco-innovation, such as to support investment in renewable energy and energy efficiency, leads to a positive outcome in terms of GDP and employment as compared to other scenarios. These findings correspond to results from other studies, such as the EmployRES study (ISI *et al.*, 2009).

3.3.2 ETR and equity implications

The issue of equity is a highly debated topic when considering the political feasibility of an ETR. Energy taxes have been found to have regressive implications as increases in environmental taxes are often thought to fall disproportionately on low-income and rural households. However, an ETR also includes the redistribution of the ETR revenues so it is of central importance to assess the *net* distributional effects of an ETR.

Energy taxes are not always regressive. Studies have demonstrated that 'taxes on household energy tend to be clearly regressive, while transport-related taxes have mixed distributional results' (Blobel *et al.*, 2009, p.15). Furthermore, country- and region-specific factors must also be kept in mind when assessing the distributional implications of environmental taxes.

The distributional impacts of ETR are not only limited to the effect on incomes, but also have effects on the environment in which different socio-economic groups live. The burden of much existing environmental pollution and degradation often falls more onto the poorer parts of a population, and they also frequently have less access to green environments, so the expected improvements to the environment from environmental taxes, as well as the effect of reducing some labour taxes as part of an ETR, especially if targeted on the less skilled, will help to redress the greater impacts of some environmental taxes on poorer households.

The large-scale ETR aiming to achieve the EU's GHG reduction target of 20% by 2020 – as modelled in the PETRE project – was the basis of the analysis of the net distributional effects of an ETR. The implications of the ETR[8] are shown in Figure 2.3.

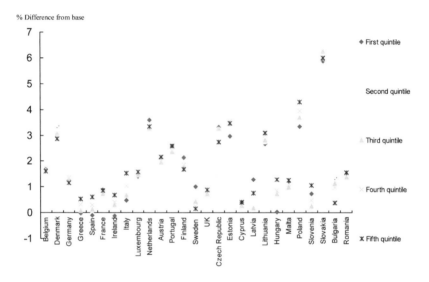

Source: Blobel *et al.*, 2009

Figure 2.3 Change in household income by quintile, S1L, 2020

The modelling results showed that, at an EU aggregate level, the ETR would generally create a positive change in real incomes for all socio-economic groups and that only a limited number of households in some EU member states would be facing a lower income as a consequence of the ETR. The results also show considerable variation in the changes in the income between countries.

It is evident that there is much more difference in the changes in income between countries than within countries. Whilst the changes in income are usually fairly similar for each income group within one country, the differences between countries are often large. For instance, whilst all income quintiles within the UK experience changes in income of around 1%, in Slovakia the changes in income are reported to be around 6.5%. (Blobel *et al.*, 2009, p.34)

There are different reasons for these variations in income including the different patterns of source of income and the different expenditure patterns in each household group between countries.

This research indicated 'that ETR could result in increases in real incomes for all groups and hence encourage employment, supporting the case for future ETR in the EU. In most countries, and at the aggregate EU level, the impacts were not found to be regressive across income groups' (Blobel *et al.*, 2009, p. 36).

Regressive effects for particular groups can further be mitigated by various redistribution and compensation mechanisms. In addition, it has to be noted that the recycling mechanism used in this modelling exercise can be described as a rather crude one as the recycling was done by reducing income taxes. The poorer part of the society does not have to pay any income taxes as their income may be below the tax threshold, so they do not benefit from the recycling measure modelled. However, other recycling measures could be implemented to provide net benefits to poorer group, e.g. by targeted benefit payments which can be a more effective way of addressing the distributional effects of ETR (OECD, 2010a).

3.4   The UK Green Fiscal Commission

The objective of the Green Fiscal Commission (GFC) was to prepare the ground for a significant programme of green fiscal reform in the UK, in terms of both assembling the evidence base for such a reform, and raising stakeholder and public awareness of it. The key findings of the UK GFC are presented in Box 2.1 (GFC, 2009, p. 6):[9]

BOX 2.1 KEY MESSAGES OF THE UK GREEN FISCAL COMMISSION

Environmental taxes work: numerous studies, including those of the Green Fiscal Commission, have shown that green taxes are effective in reducing the environmental impacts on which they are targeted.

Environmental taxes can raise stable revenues: some environmental taxes, like fuel duty, have been raising sizeable revenues for years. Raising them significantly would therefore both achieve environmental improvements and allow other taxes to be lower than they would otherwise need to be.

The UK's 2020 greenhouse gas targets could be met through green fiscal reform: the economic implications of doing so would be broadly neutral, and the green fiscal reform policy approach would increase employment.

The impacts of green fiscal reform on competitiveness can be mitigated: relatively few economic sectors would face serious challenges to their competitiveness from green fiscal reform, and there are a number of ways in which these concerns can be addressed.

Green fiscal reform emerges as a crucial policy to get the UK on a low-carbon trajectory; help develop the new industries that will both keep it there and provide competitive advantage for the UK in the future; and contribute to restoring UK fiscal stability after the recession. It is a key to future environmental sustainability and low-carbon prosperity.

4. A SUMMARY: THE POTENTIAL OF ENVIRONMENTAL TAXES AND ETR IN THE CURRENT POLICY DEBATE

Climate change, biodiversity loss, ecosystem degradation, growing material resource scarcity, as well as national budget deficits and an increasingly ageing population are current challenges facing the EU. At the same time, there is increased understanding of the interlinkages between many environmental, economic and social problems, pointing to the cost-effectiveness of integrated packages of policy measures (EEA, 2010).

Public finances deteriorated during the financial and economic crises of recent years. Many attempts are being made to consolidate the large public deficits which are often the outcome of unprecedented fiscal interventions and fiscal stimulus packages (EC, 2010a). Therefore it is crucial to implement policy instruments, such as environmental taxes and an ETR, which can address these challenges. The EC observed that environmental taxes are likely to play a key role in the near future and in particular in the context of the current debate of revenue shortfalls (EC, 2010a).

Fiscal consolidation strategies can be implemented at the spending and at the revenue side of the public budget. Some countries, such as Spain,

Greece, Ireland and the UK, have implemented large austerity programmes by reducing spending. However, the mix of instruments achieving budget consolidation depends on country-specific factors. A recent IMF report (2010) stated that developed countries will probably rely heavily on spending cuts (austerity programmes). What remains crucial is that all programmes must consider the significant challenges lying ahead in relation to the ageing population whereby the labour tax base will be shrinking whilst both the lifetime consumption tax base and public expenditure on the elderly, will be rising. This is also the case when fiscal consolidation will be done via the revenue side: one of the features discussed by IMF is to strengthen broad-based taxes by listing an increase in excise tax rates and by 'introducing (and capturing revenues from) efficient carbon pricing in the United States and Europe' (IMF, 2010, p. 4). It is furthermore stated in this report that:

> *Broad-based consumption taxes and property taxes are less harmful to growth than income taxes.* Taxing consumption is equivalent to taxing accumulated assets and labour income: so it falls partly on a completely inelastic base – previously existing assets – and partly on a base less internationally mobile than capital income. (IMF, 2010, p. 31)

This is confirmed by an EC analysis (EC, 2010a).[10] A further advantage of environmental taxes is the fact that they can stimulate the development and diffusion of new technologies and practices.

ETR is therefore a valuable measure for dealing with both short term budgetary imbalances and spurring the transition towards a green economy, as recently acknowledged by the OECD (2010b) and by the Europe 2020 strategy 'A European strategy for smart, sustainable and inclusive growth':

> Where taxes may have to rise, this should, where possible, be done in conjunction with making the tax systems more 'growth-friendly'. For example, raising taxes on labour, as has occurred in the past at great costs to jobs, should be avoided. Rather Member States should seek to shift the tax burden from labour to energy and environmental taxes as part of a 'greening' of taxation systems. (EC, 2010b, p.24)

In 1993 the Delors *White Paper on Growth, Competitiveness and Employment* observed that the European economy was characterised by 'an insufficient use of labour resources and an excessive use of environmental resources' (EC, 1993). It is concluded:

Finally, if the double challenge of unemployment/environmental pollution is to be addressed, a swap can be envisaged between reducing labour costs through increased pollution charges. (EC, 1993)

The time has come to act on this observation.

NOTES

1. See for a discussion of the two concepts: EEA, 2005 and OECD, 2006.
2. The findings of the project are published in Andersen and Ekins, 2009.
3. See for a more detailed discussion on the double dividend theory: Pearce, 1991, and Repetto *et al.*, 1992.
4. The findings of the project are published in Ekins and Speck, 2011.
5. See for a detailed discussion of the models: Barker *et al.*, 2011.
6. See for more information on the GRAM model: Giljum *et al.*, 2011.
7. A total of six scenarios were modelled – the two scenarios discussed in this chapter are:
    – Scenario HS1: ETR with revenue recycling (high oil price) designed to meet unilateral EU 2020 GHG target (20% reduction from 1990 levels)
    – Scenario HS2: ETR with revenue recycling (high oil price) designed to meet unilateral EU 2020 GHG target, with 10% of revenues spent on eco-innovation measures.
    A complete discussion of the models, scenarios and results can be found in *Barker et al.*, 2011.
8. A detailed analysis and underlying assumptions of the scenarios can be found in Blobel et al., 2009.
9. The final report as well as a range of briefing papers can be downloaded from the Green Fiscal Commission website http://www.greenfiscalcommission.org.uk.
10. See also Johannson *et al.*, 2008 and Koske, 2010.

REFERENCES

Andersen, M.S. and P. Ekins (eds.) (2009), *Carbon-Energy Taxation: Lessons from Europe*, Oxford University Press, Oxford.
Barker T., S. Junankar, H. Pollitt *et al.* (2009a), 'The Effects of Environmental Tax Reform on International Competitiveness in the European Union: Modelling with E3ME', in: M.S. Andersen and P. Ekins (eds.), *Carbon-Energy Taxation: Lessons from Europe*, Oxford University Press, Oxford, 147–214.
Barker T., Dagoumas A. and J. Rubin (2009b), 'The Macroeconomic Rebound Effect and the World Economy', *Energy Efficiency*, **2**, 411–27.
Barker T, C. Lutz, B. Meyer *et al.* (2011), 'Modelling an ETR for Europe', in: P. Ekins and S. Speck (eds.), *Environmental Tax Reform: A Policy For Green Growth*, Oxford University Press, Oxford.
Bassi S., P. ten Brink, M. Pallemarts *et al.* (2009), *Feasibility of Implementing a Radical ETR and its Acceptance*. Final Report of the 'Study on tax reform in Europe over the next decades: implementation for the environment, for eco-innovation and for household distribution' commissioned by the European Environment Agency (EEA), Copenhagen.
Blobel D., H. Pollitt, T. Drosdowski *et al.* (2009), *Distributional Implications: Literature Review, Modelling Results of ETR – EU-27 and Modelling Results of*

ETR – Germany. Final Report of the 'Study on tax reform in Europe over the next decades: implementation for the environment, for eco-innovation and for household distribution' commissioned by the European Environment Agency (EEA), Copenhagen.

Ekins P. and S. Speck (1999), 'Competitiveness and Exemptions from Environmental Taxes in Europe', *Environmental and Resource Economics,* **13**, 369–96.

Ekins P. and S. Speck (2008), 'Environmental Tax Reform in Europe: Energy Tax Rates and Competitiveness', in: L. Kaiser, J. Milne and N. Chalifour (eds.), *Critical Issues in Environmental Taxation: Volume V,* Oxford University Press, Oxford, 77–105.

Ekins P. and S. Speck (eds.) (2011), *Environmental Tax Reform: A Policy For Green Growth*, Oxford University Press, Oxford.

ESCAP (2008), *Greening Growth in Asia and Pacific*, Bangkok.

European Commission (EC) (1993), *Commission White Paper on Growth, Competitiveness and Employment*, COM(93)700 final, Brussels.

European Commission (EC) (2010a), *Monitoring tax revenues and tax reforms in EU Member States 2010*, European Economy 6/2010, DG for Economic and Financial Affairs and DG for Taxation and Customs Union, Brussels.

European Commission (EC) (2010b), *A European strategy for smart, sustainable and inclusive growth*, COM(2010)2020final, Brussels.

European Environment Agency (EEA) (1996), *Environmental Taxes – Implementation and Environmental Effectiveness*, Copenhagen.

European Environment Agency (EEA) (2000), *Environmental Taxes: Recent Developments in Tools for Integration*, Copenhagen.

European Environment Agency (EEA) (2005), *Market-based Instruments for Environmental Policy in Europe*, EEA Technical Report No8/2005, Copenhagen, Denmark.

European Environment Agency (EEA) (2010), *The European Environment: State and Outlook 2010 – A Synthesis,* Copenhagen.

Eurostat (2010), *Taxation Trends in the European Union Data for the EU Member States, Iceland and Norway, 2010 edition*, a report prepared by Eurostat and European Commission, Taxation and customs union Luxembourg, Publication Office of the European Union.

Fraunhofer ISI, ECOFYS, Energy Economics Group, Rütter + Partner, SEURECO & LEI (ISI *et al.*) (2009), *The Economic Impact of Renewable Energy Policy on Economic Growth and Employment in the European Union*, Brussels.

Gehr U., C. Lutz and R. Salmons (2009), *Eco-Innovation: Literature Review on Eco-innovation and ETR and Modelling of ETR with GINFORS*. Final Report of the 'Study on tax reform in Europe over the next decades: implementation for the environment, for eco-innovation and for household distribution' commissioned by the European Environment Agency (EEA), Copenhagen.

Giljum S., C. Lutz, A. Jungnitz *et al.* (2011), 'European Resource Use and Resource Productivity in a Global Context', in: P. Ekins and S. Speck (eds.), *Environmental Tax Reform: A Policy For Green Growth*, Oxford University Press, Oxford.

Green Fiscal Commission (2009), *The Case for Green Fiscal Reform Final Report of the UK Green Fiscal Commission*, London.

International Monetary Fund (IMF) (2010), *From Stimulus to Consolidation: Revenue and Expenditure in Advanced and Emerging Economies*, paper prepared by the Fiscal Affairs Department, April 30, 2010, Washington.

Johannson A., C. Heady, J. Arnhold *et al.* (2008), *Tax and Economic Growth*, OECD Economics Department Working Paper No.620, ECO/WKP(2008)28, Paris.

Koske I. (2010), *After the Crisis: Bringing German Public Finances back to a Sustainable Path*, OECD Economics Department Working Paper No.620, ECO/WKP(2010)22, Paris.

Lutz, C and B. Meyer (2009), 'Environmental and Economic Effects of Post-Kyoto Carbon Regimes. Results of Simulations with the Global Model GINFORS', *Energy Policy*, **37**, 1758–66.

Organisation for Economic Co-operation and Development (OECD) (2001), *Environmentally Related Taxes in OECD Countries: Issues and Strategies,* Paris.

Organisation for Economic Co-operation and Development (OECD) (2005), *Environmental Fiscal Reform for Poverty Reduction*, DAC Guidelines and Reference Series, DAC Guidelines and Reference Series, Paris.

Organisation for Economic Co-operation and Development (OECD) (2006), *The Political Economy of Environmentally Related Taxes*, Paris.

Organisation for Economic Co-operation and Development (OECD) (2010a), *Taxation, Innovation and the Environment*, Paris.

Organisation for Economic Co-operation and Development (OECD) (2010b), *Interim Report of the Green Growth Strategy: Implementing our commitment for a sustainable future*, Meeting at the OECD Council at Ministerial Level, 27–28 May 2010, Paris, G/MIN(2010)5.

Pearce D. (1991), 'The Role of Carbon Taxes in Adjusting to Global Warming', *Economic Journal*, **101**, 938–48.

Repetto R., R.C. Dower, R. Jenkins *et al.* (1992), *Green Fees: How a Tax Shift Can Work for the Environment and the Economy,* World Resources Institute, Washington, USA.

Sorrell, S. (2007), *The Rebound Effect: An Assessment of the Evidence for Economy-wide Energy Savings from Improved Energy Efficiency*, London: UK Energy Research Centre.

Speck S. and J. Jilkova (2009), 'Design of Environmental Tax Reforms in Europe', in: M.S. Andersen and P. Ekins (eds.), *Carbon-Energy Taxation: Lessons from Europe*, Oxford University Press, Oxford, 24–52.

World Bank (2005), *Environmental Fiscal Reform. What Should be Done and How to Achieve it*, Washington DC.

3. Comparisons between the cap and trade system and carbon taxation: is the USA ready for a carbon tax?

Paul J. Lee, Rahmat O. Tavallali, Hai Sook Kwon and John T. Geekie

INTRODUCTION

The current global environmental crisis and the downturn in the US economy are two very large issues which weigh heavily on the minds of Americans. In particular, carbon emissions and the dangerous burden that global warming poses for the future are problems which must be addressed as quickly as possible. The cap and trade model was featured in US President Barack Obama's first budget as a way to tackle the environmental devastation the world faces. However, the severe problems with the economy, protests from critics, and the complex nature of implementing such a system prevented it from being established in a timely manner (Broder, 2009). For some time, Congress was re-evaluating the possibility of employing some form of the cap and trade system (*The New York Times*, March 2010). However, there is a need to examine whether or not that particular model is effective enough to be implemented on its own, or whether it should be preceded by or combined with another method. This paper will discuss the option of using a taxation method to precede the cap and trade system, which could takeover in the long term, in an effort to take immediate action which will reduce carbon emissions.

Carbon taxation could be an effective method in pushing companies towards reduction of carbon usage, which would result in reform on a broader scale. It can even be immediately applied to the individual consumer. Ideally, carbon taxation need not be employed as an additional tax; rather, it could be used as a shifting of funds. Instead of only a penalty or punishment to prevent unwanted behavior, the revenue could be used as incentives to encourage Americans in an effort to modify the use of fossil fuels. In this manner, carbon taxation may be able to have a substantial and positive influence on the daily lives of all Americans. This chapter will look at the strengths and weaknesses of the carbon taxation system, and it will also compare and contrast it to the cap and trade system method. The chapter will also discuss tax incentive options to use funds in productive ways to encourage a reduction in usage of resources and economic stability. Furthermore, the chapter will address using tax incentives to promote the usage of cleaner fuels.

THE NECESSITY FOR ACTION

Global warming is undeniable at this point in time; scientists across the board agree that people are causing an enormous greenhouse effect due to the amount of carbon dioxide emissions (UNFCCC, 2009). In 2007, the Intergovernmental Panel on Climate Change (IPCC) issued a report stating that there is a 90 percent likelihood that people are creating global warming (IPCC, 2007), suggesting that any of the ecological changes that are occurring are unnatural. When examining data from 1850 to 2009, the greatest rise of carbon dioxide emissions was within the past 30 years, which coincides to the significant increases in industrialization and globalization (US EPA, 2010). Indeed, it is projected that without any effort to reduce greenhouse gases, there will be a rise in the global average temperature from 2° Fahrenheit to 11.5° Fahrenheit by 2100 (Env. Defense Fund, 2009). While this might not sound like a very significant change, it is actually an enormous increase which will result in unforeseeable consequences.

With such alarming scientific data, there is great pressure to compel the current government to implement an effective strategy immediately. There is little time to work on an intricate alternative fuel source, let alone sufficient time to develop a way to clean out the carbon dioxide that has already been emitted. Companies have been striving to make an effective alternative fuel source to fossil fuels and some fads, feeble attempts at 'going green,' have come and gone. Ethanol, corn, hydrogen power, solar power, electricity and other ideas have been experimented with, yet there is not a power source that seems to be able to wean the United States from the overconsumption of fossil fuels. Therefore, it is imperative to reduce the current usage as soon as possible and look forward to establishing a long-term solution as well.

While there were numerous attempts to produce a viable alternative fuel, there have also been innovations in using economic policies to further influence the reduction of carbon emissions. The Kyoto Protocol, established in 1997, was adopted by 37 countries, and it called for a mandatory reduction of greenhouse emissions by at least 5.2 percent starting in 2008 (Lee, 2010). Unfortunately, the United States chose not to participate in the Kyoto Protocol and has continued to generate the pollution that has contributed to so many environmental problems. As the primary contributor of greenhouse pollution in the world, the United States as a whole has not been particularly eager to adopt a policy to penalize companies or individuals for their carbon emissions. However, market-based approaches, such as cap and trade or carbon taxation, might be able to tamper down the level of carbon emissions while simultaneously stimulating the economy.

Cap and Trade

Currently, the US government is looking at the option of the cap and trade system. The premise of the cap and trade system is simple. The 'cap' is a government-allotted amount of emissions each company will be allowed to produce. The purpose of this cap is to minimize the overall carbon emissions to a safer level than currently permitted. The 'trade' refers to the allowance for those companies that do not use all of their allotted emission to 'trade' their credits to other companies that need it for profit (*The New York Times*, July 2010). It has been shown that a well-established cap and trade system can be effective in decreasing the costs of emissions in comparison to normal regulatory methods (Stavins, 2003).

Rather than primarily being a government-controlled direct regulation, the cap and trade system can instead be indirectly regulated by the market. The cap can serve as an accountability system amongst companies, while trading allows each company to have the independence and flexibility to function. The idea is to use this system to balance and encourage flexibility, while keeping emissions down and preventing further environmental damage. However, this system has not yet been proven to work. The European Emissions Trading Scheme (Kruger, 2007) is by far the most extensive attempt at the cap and trade to date. It has passed through the first phase (from 2005 to 2007), where an attempt to trade for carbon dioxide emissions was criticized because the initial cap was set so high 'due to a vague, decentralized process of setting the caps and determining allowances by individual countries' (Kruger, 2007), no country was penalized and no companies had to moderate their carbon pollution (Convery, 2007). The European countries are currently in the midst of the second phase, which includes an attempt to regulate other greenhouse gases, but it is uncertain as to whether it will be effective.

The greatest fear is that economically the cap and trade system might not work because it will not have a chance to prove itself to be a worthy solution to the global warming problem. As the European Emissions Trading Scheme has shown, during the first two years a cap and trade system was implemented, it was with sub-par results (Convery, 2007). The reality is that expecting a smoothly working cap and trade system within any immediate time frame is foolish. If it falters towards the way that Europe has experienced, with a cap set too high, the United States will have a poorly organized system that will neither reduce the carbon emissions nor change the behaviors of participating companies (Kruger, 2007).

On the other hand, if the cap is set too stringently, as feared by critics (mainly large companies with high carbon production), the system will become mainly punitive. As companies are required to trade for more credits because of their overusage of the limited allowances, it stands to reason that they will have to require greater payment from consumers, which significantly increases the burden on a public that is already struggling with the floundering economy (*The New York Times*, July 2010).

The time to establish a cap and trade system is also important to consider. If prematurely implemented, the cap and trade system has the potential to be very ineffective and thereby harmful. If ineffective in lowering carbon emissions, the cap and trade system can delay appropriate action against carbon pollution and contribute more carbon emissions to the current burden, creating further environmental devastation. The high-stakes risks of the cap and trade system are especially difficult to adopt as environmental policy during the current downturn of the economic climate. The fear of failure is currently exemplified in other countries' attempts at using a similar system. Australia is trying to establish a cap and trade system, but Julia Gillard, the current prime minister, is delaying the implementation of this system until previous commitments can be completed (Australian Government, 2010). It is of utmost importance to prevent future resistance of a promising policy by first closely examining and then establishing the appropriate method to solve this problem (Repetto, 2007).

Stavins suggests that a cap and trade system would be most successful as long as it fulfilled several criteria: a gradually increasing trajectory of emissions reductions over time, tradable allowances, upstream regulation, mechanisms to reduce uncertainty, phasing in full auctioning of cap and trade credits, and linkages to other cap and trade systems around the world (Stavins, 2003). The potential for the cap and trade system to work as environmental taxation policy is dependent upon how earnestly the nation is committed to working in a flexible and creative manner, both within its own national framework and on a global scale. While this system would be the ideal to implement in the long term, something must be done immediately to stop the prevailing global harm as soon as possible (UNFCCC, 2009).

Carbon Tax

An alternative solution is carbon taxation. As with cap and trade, the measurement of each ton of carbon emission per unit of energy used, the British Thermal unit is the focus of the policy. The emission is measured per company by area, again, similar to the cap and trade. However, rather than using the number of units as a tradable item, carbon taxation uses the measurements as an indicator of how much the producers of carbon emissions should be taxed. There are several benefits with a carbon tax: it is quicker to implement; it encourages the reduction of carbon emissions; and it is enforceable on a smaller scale as well.

A carbon tax is very simple; it would require minimal new bureaucracy and could be implemented in a relatively short time span. In 1990, Finland was the first country to enact a carbon tax (Snowdon, 2008). At first it was solely a tax on carbon content, but it was later changed to a combined carbon and energy tax. As other European countries adopted this tax policy, taxes were also applied to non-industrial consumers for electricity and renewable fuels, which led to an increase in use of renewable fuels for heating and electricity (Snowdon, 2008).

In addition, a carbon tax encourages the reduction of carbon emissions, as companies would strive to avoid incurring large sums of taxes. The carbon tax would be an initiative for companies which currently over-produce carbon emissions to improve their habits or else pay for the damage.

Another important point about the carbon tax program is that it has the potential to elicit the participation of every person and every company much more easily than a cap and trade system would. The consumer could also be directly taxed depending upon the car that he or she drives and the fuels used, for example. Obviously, this could be a very useful method to influence the way that consumers begin to view carbon taxation. Behavior modification of the US population by carbon taxation is potentially one of the strongest benefits a carbon taxation policy can offer; individuals can make a significant impact on decreasing the carbon emissions. The production and sale of hybrid cars would become more popular, as they both emit less carbon and use alternative fuels. Therefore, carbon taxation can be the impetus for further innovations in itself and can be used positively for both companies and individuals.

It is important to acknowledge that a carbon tax provides a strong potential for stimulation of the economy; a carbon tax policy might further encourage public participation in decreasing carbon emissions by using the revenue to provide additional incentives to those who perform better by producing less carbon emissions. The government can avoid a straight addition of more taxes to US citizens. Rather, carbon taxation provides tax incentives to those who are already making good choices in the reduction of greenhouse gases and to those who are willing to become innovators.

TAX INCENTIVES FOR THE UNITED STATES

Using a carbon tax would be an effective method to reduce the emission of carbon, but the United States could further benefit from carbon taxation if the method included distribution of the funds across the nation with the intention of stimulating the economy. In the UK, climate change taxes are used to pay for cuts in employers' national insurance contributions and for additional support in energy efficiency and renewable energy (Snowdon, 2008). An even more powerful prospect would be to tax those who emit greater amounts of carbon and use the tax money to distribute amongst those who are taking steps to conserve fuels. There are two items – automobiles and fuels – which could really contribute to the success of a carbon tax both as a method of reducing carbon damage and as an incentive for consumers.

One of the most practical applications of the carbon tax to the general population would be to more heavily tax those with fuel-inefficient automobiles. This would be aimed at influencing the public's behavior in purchasing more efficient cars. A carbon tax targeted at automobiles could be further utilized to subsidize the costs of more fuel-efficient cars and hybrids for other ecologically minded consumers. The method of the provision of tax incentives for hybrids could take the form of either lowering the initial cost of the vehicle or providing a rebate, both of which are easy and enforceable tasks. Currently, there are already certain hybrid or clean fuel burning cars that qualify for tax incentives ('Fed. Tax', 2010). It is undeniable that the tax incentives and rebates have the possibility to exert a strong influence on the public. It has been shown that US consumers are already mindful about gasoline prices and the per mile cost of vehicles in making their purchase decisions (Bento, 2009). In addition, there is a positive correlation to the public responding to tax incentives to buy hybrid cars (Lazzari, 2006). A carbon tax applied to automobiles would create a strong motivation for the public to start looking into investing in environmentally friendly automobiles and to decrease the popularity of those which currently create so much pollution.

The adoption of carbon taxation on gasoline prices can also influence behavior modification for individual consumers. It has been shown that government policy which targets both emissions pricing and research and development (R&D) is more cost-effective than emissions pricing alone (Furman, 2007). If this is the case, it is possible that the very revenue which would decrease the unwanted pollution could very effectively be the funds which could even kick-start the alternative fuel research that is so necessary for the United States. With a well-funded and better-supported field of research, the race to develop a viable alternative fuel might be successful more quickly. Carbon taxation of gasoline prices can be effective in lowering the current dangerous levels of emission while also encouraging and developing cleaner future alternative fuels.

Cap and Trade vs. Carbon Taxation

Both cap and trade and carbon taxation systems are market-based policies which can make a large impact on the US economy. However, it is with great care that each of these policies should be regarded. In comparing the two methods, the important factors are: how efficiently the policy can be implemented, how well the policy can be managed, and how the policy influences both the environment and the economy in the future.

The current environmental problems require immediate reparation. Therefore, any environmental policy should be very efficient and make a significant impact on the current damages. The experiences of other countries which have implemented either of the policies can inform us on which approach might be more useful for the immediate future. While it took British Colombia a short five months to establish a working taxation system, it took five years for the US' north-eastern region to figure out a workable cap and trade system (Duff, 2008). The significant difference in initiation time might be reduced with a more organized cap and trade system; however, even the carbon tax model used by British Columbia can also be improved and established more quickly. With the urgency that is stressed by President Obama and other world leaders, the carbon tax is more convincing in the area of efficiency of initiation time.

Aside from the pressure of time, a carbon tax system might also prove to be more manageable than a cap and trade system. If each company or consumer is held accountable for emissions, it would be more straightforward to tax and obtain the taxes. Tax integration can become a very streamlined method of merging personal and corporate taxes for the advantage of not only the market but also of the environment (Metcalf,

2007). On the other hand, a cap and trade system is meant to manipulate and bargain for credits, by nature. The cap and trade system can shift blame around and around, until ultimately, a few of the losers of the game might be massively penalized.

Some critics may be reluctant to agree to carbon taxation because they might be afraid of large sums of punishing taxes due. However, this has not been shown to be the case. In Boulder, Colorado, the tax is costing each household a mere $1.33 (Carbontax.org, 2010). While a carbon tax policy applied on a national level might not be quite as low as that of Boulder, it is unlikely that a carbon tax would be prohibitive on US citizens.

In addition, it is more likely that the carbon tax model would promote creative and effective alternative energy research. There is also the potential for the taxation of larger companies to be a positive influence on their provision of goods in a more ecological way as well. While a cap and trade model is a process of exchange between the large companies, there might not be any particular force on behavior modification on the public. Cap and trade might become a very overbearing and possibly oppressive force if a company overruns its cap and needs to raise revenues from the consumer to have more to trade. In this way, any modifications to society made by cap and trade would be a penalty. On the other hand, a carbon tax policy can be applied for a very direct and very comprehensive target and achieve both short-term and long-term goals in carbon emission reduction. Furthermore, the carbon tax system allows for further improvements at the community level. While a carbon tax policy can be adjusted to alleviate some of the socio-economic inequalities in low-income households, cap and trade systems do not have a mechanism to decrease the additional burden of the high-energy prices on deprived populations. Instead, cap and trade systems increase the inequality in burden-to-income ratios between high- and low-income populations (Aldy, 2008). In 2008, the Congressional Budget Office issued a statement stating that the 'net benefits (benefits minus costs) of a tax could be roughly five times greater than the net benefits of an inflexible cap' (Congressional Budget Office, 2008). Carbon taxing may not fix the pre-existing environmental damage, but it is a step in the right direction toward decreasing pollution and raising awareness of the importance of the issue of carbon emissions.

CONCLUSION

There are strong advocates for both sides of these issues. However, it appears that neither method should be expected to be able to stand alone. Although the cap and trade system would help in entirely limiting the usage of carbon emitting fuels, it leaves significant room for error as a system to be implemented from the very beginning of this quest to lower carbon emissions. On the other hand, whereas companies may initially resist carbon taxation, it would provide them with a tangible burden of the cost that pollution incurs on the environment and encourage them to lessen their usage and improve the environment. In addition, the carbon tax has real potential to benefit the US economy as well, if the revenue were allocated to focus on tax incentives for fuel-efficient vehicles and alternative fuel research. Therefore, both cap and trade and carbon taxation can be implemented together to create a more effective and timely solution.

REFERENCES

Aldy, Joseph Edward, Ley, Eduardo and Parry, Ian (2008), 'A Tax-Based Approach to Slowing Global Climate Change', *National Tax Journal*, Vol. LXI, No. 3.

Australian Government, Department of Climate Change and Energy Efficiency (2010), 'Carbon Pollution Reduction Scheme', available at www.climatechange.gov.au, accessed 30 July 2010.

Bento, Antonio M., Goulder, Lawrence H., Jacobsen, Mark R., *et al.* (2009), 'Distributional and Efficiency Impacts of Increased US Gasoline Taxes' *The American Economic Review*, 99.3, 667–99.

Broder, John M. (2009), 'From a Theory to a Consensus on Emissions', *The New York Times*, 17 May 2009: A1. Print.

Carbon Tax Center, The (2011), 'Where Carbon is Taxed', The Carbon Tax Center, available at www.carbontax.org, updated 25 July 2011, accessed 31 July 2011.

Congressional Budget Office (2008), 'Policy Options for Reducing CO Emissions', p. IX, available at www.cbo.gov/ftpdocs/89xx/doc8934/02–12-Carbon.pdf.

Convery, Frank and Redmond, Luke (2007), 'Market and Price Developments in the European Union Emissions Trading Scheme', *Review of Environmental Economics and Policy*, 1, 66–87.

Duff, David G. (2008–09), 'Carbon Taxation in British Columbia', *Vermont Journal of Environmental Law*, 10, 87, 87–107.

Environmental Defense Fund (2009), 'The Basics of Global Warming: The Green-house Effect', available at www.Edf.org, EDF, accessed 29 July 2010.

'Federal Tax Incentives (United States)' (2010), *Hybrid Cars*, www.hybridcars.com, 25 February 2010, accessed 29 July 2010.

Furman, Jason, Bordoff, Jason E., Deshpande, Manasi, *et al.* (2007), 'An Economic Strategy to Address Climate Change and Promote Energy Security', *The Hamilton Project, Strategy Paper*. Washington, DC, The Brookings Institution, October 2007.

Intergovernmental Panel on Climate Change (IPCC) (2007), 'Climate Change 2007: The Physical Scientific Basis', *Summary for Policymakers 2007*.

Kruger, Joseph, Oates, Wallace E., and Pizer William A. (2007), 'Decentralization in the EU Emissions Trading Scheme and Lessons for Global Policy', *Review of Environmental Economics and Policy,* 1, 112–33.

Lazzari, Salvatore, 'Tax Credits for Hybrid Vehicles', *Congressional Research Service: CRS Report RS22558*. 20 Dec. 2006.

Lee, Jee Hoon (2010), 'Post-Kyoto Protocol and Emissions Trading', *SERI Quarterly*, 23–27.

Metcalf, Gilbert E. (2007), 'Corporate Tax Reform: Paying the Bills with a Carbon Tax', *Public Finance Review,* 35 (3), 440–59.

New York Times, The (2010), 'Cap and Trade', *Times Topics – The New York Times*, available at www. topics.nytimes.com. Updated: 26 March 2010, accessed 31 July 2010.

Repetto, Robert (2007), 'National Climate Policy: Choosing the Right Architecture', *Yale School of Forestry and Environmental Studies*, June 2007.

Snowdon, Catherine (2008), 'Taxing the environment', *International Tax Review*, 19.9.

Stavins, Robert N. (2003), 'Experience with Market-Based Environmental Policy Instruments', in Karl-Göran Mäler and Jeffrey Vincent (eds), *Handbook of Environmental Economics*, Volume I, Amsterdam: Elsevier Science, 355–435.

UNFCCC Negotiations (2009), 'G8+5 Academies' joint statement: Climate change and the transformation of energy technologies for a low carbon future', G8+5 Academies: May 2009, PDF file, available at www.nationalacademies.org/includes/G8+5energy-climate09.pdf.

United States Environmental Protection Agency (2010), 'Climate Change Science Facts', April 2010, PDF file, available at www.epa.gov/climatechange/downloads/Climate_Change_Science_Facts.pdf.

PART II

Environmental Taxation Policy
Considerations

4. Innovative taxation strategies supporting climate change resilience

Rolf H. Weber

I. INTRODUCTION

Currently no globally accepted carbon price scheme exists; hence regulations concerning emissions reductions are mainly governed by national laws (Weber and Kaufmann, 2011). Since governmental measures designed to improve climate conditions, such as incentive-driven carbon taxes or cap-and-trade-regimes (other measures are, for example, standardization rules or prohibition legislation: Avi-Yonah and Uhlmann, 2009, p. 22 *et seq.*), may differ between countries, the problem could arise that these differences lead to anti-competitive or protectionist effects, because producers of goods may have to face higher costs, if they are subject to stringent environmental provisions.

In light of this fact, the risk is immanent that carbon intensive industries relocate to countries having less strict provisions on carbon emissions. This phenomenon is generally referred to as 'carbon leakage'. Furthermore, 'carbon heavens' could pose a threat to the effectiveness of globally agreed carbon reduction measures (WTO, 2009, pp. 98–100, WTO, 2009a).

To combat the negative effects of climate change trough reduction of carbon emissions, national governments rely mostly on 'traditional' tax instruments such as carbon taxes and cap-and-trade regimes. Both instruments are designed to modify the behavior of the market participants by sending price signals to the market. These instruments may have an effect on a national scale; however, they do not encompass entities producing abroad and, therefore, the introduction of new environmental taxation strategies must be considered.

The concept of creating a powerful incentive for the producers of negative externalities to reduce their output was developed in the early 20th century by the distinguished economist Pigou. Stating that polluters do not account for 'social costs' imposed on others by their emissions, Pigou proposed that it should fall within the responsibility of the state to create extraordinary encouragements or extraordinary restraints to mitigate the social costs (Pigou, 1932). His concept of addressing market failure by levying taxes was subsequently further developed by several scholars and is even today still significant (Yandle, 1999, p. 8 *et seq.*).

A generally accepted definition of the term environmental tax is not yet available. The Organisation for Economic Co-operation and Development (OECD) refers to the environmental tax as a tax based on polluting emissions or on disamenities expressed by appropriate methods of measurement or on other parameters such as inputs (OECD, 1980/ Määttä, 2006, p. 15). Therefore, the environmental taxes generally encompass governmental tax policies, which are shaped in such a way that they have an effect on the environment (Xu, 2010, p. 4). These taxes can be levied on every possible polluting material; taxes on carbon emissions undoubtedly fall under this definition. (i) A carbon tax should lead to a decrease of emissions (thereby helping to combat the climate change); (ii) the state accrues revenue which can be invested in alternative energy sources; (iii) producers complying with higher environmental standards would benefit from financial advantages.

In general, environmental taxes can be divided into three major categories, namely (i) taxation on the emissions themselves, (ii) taxation on the users of certain facilities which produce emissions and (iii) taxation on goods or services that are generating pollution in the manufacturing, consumption or disposal phase (World Bank Group, 1998, p. 161). Another important distinction concerns the nature of taxes. (i) Direct taxes are paid by the individuals or entities they are imposed on. Such kind of taxes, for example, taxes on income or assets, cannot be shifted to another person or legal entity.[1] (ii) Indirect taxes, however, for example, a sales tax, are levied on an entity, but are ultimately paid by another person or entity.[2] The main objective of all types of environmental taxes is to create an incentive for every producer of harmful emissions to reduce its output of carbon dioxide in order to stay competitive (Enders and Ohl, 2005, p. 21).

II. TRADITIONAL TAX INSTRUMENTS

In order to lay the ground for the analysis of navigating international trade and tax rules in a carbon leakage preventing way and of realizing competitiveness in the market, the pros and cons of the traditional instruments are shortly discussed.

1. National Carbon Taxes

a) Principles and design

A governmental decision to levy a tax on the emission of carbon dioxide in order to restrict the effect of global warming (Stern, 2007) wants to achieve an increase in costs for the emitter of carbon dioxide (Pigou, 1932), giving an incentive to reorganize the manufacturing procedures in order to cause less pollution (OECD, 2010, p. 31 *et seq.*). So far, taxes have mostly been levied directly on the emitter of carbon dioxide or imposed on goods or services if during their production or delivery carbon dioxide was emitted (Cockfield, 2011, p. 7).

These taxes can be levied on all carbon based resources such as oil, coal and natural gas. The rate of such a tax should correspond to the marginal cost of carbon dioxide emissions; annually, the tax would then be raised proportionally to the negative effects of carbon dioxide (Avi-Yonah and Uhlmann, 2009, p. 32).The revenue raised from these taxes could not only be used to countervail the harmful impact of carbon dioxide emissions, but also to fund alternative energy sources (Sandmo, 2009, OECD, 2010, p. 143/44).

The major advantage of a carbon tax is its simplicity even if it is not easy to identify the appropriate level of the tax (OECD, 2010, pp. 15, 96 *et seq.*, 138 *et seq.*). Revenue, generated by taxes, could be substantial (Avi-Yonah and Uhlmann, 2009, p. 40) and successfully be invested in alternative energies attempting to mitigate the disadvantages from pollution as first results about the effectiveness of environmentally related taxation on innovation show (OECD, 2010, p. 63 *et seq.*).

The concept of a tax on carbon dioxide also offers the producers of emissions cost certainty because the cost is the amount of the tax levied; therefore, an enterprise possesses the opportunity to assess prospective costs for their carbon emissions (Avi-Yonah and Uhlmann, 2009, p. 42). Property designed taxes are usually transparent as it is clear on what goods the taxes apply and which polluters (if applicable) are exempted (OECD, 2010, p. 137). Furthermore, due to the economic signalling effect the carbon tax will improve economic efficiency and hence a reduction of greenhouse gas emissions will be realized because the producers are forced to bear the 'damages' of their actions (Duff, 2003, p. 2069).

b) Shortcomings

Even if the concept of a carbon tax entails many positive aspects, it is afflicted with some substantial disadvantages. Firstly, the marginal social costs of carbon dioxide emissions on which the tax rate is based are extraordinarily difficult to assess (Cockfield, 2011, p. 7).

Secondly, there is uncertainty as to whether the consumers actually would purchase less carbon dioxide-intense products pressuring the producers to develop less environmentally harmful goods, since the elasticity of demand of the concerned products (so-called 'benefit uncertainty') (Cockfield, 2011, p. 8). In such case (and if there is no border tax adjustment) the tax has to be raised in order to achieve the intended environmental goals which may not be politically feasible (Avi-Yonah and Uhlmann, 2009, p. 46).

Thirdly, the implementation of a tax on carbon dioxide will presumably have to face bitter opposition not only in politics but also from the lobby of the industries, consequently leading to a lower tax base and, therefore, impairing the possibility of an environmental benefit (Avi-Yonah and Uhlmann, 2009, p. 47 *et seq.*).

Lastly, the carbon tax only shows an effect on a national scale, thus potentially leading to competitive disadvantages for the domestic industries (Weber and Kaufmann. 2011).

2.   Cap-and-Trade Regimes

a)   Principles and design

At present the most favoured carbon-pricing mechanism is the cap-and-trade regime. The European Union has implemented a trading scheme and in the United States the establishment of such system is scheduled to be incorporated in the 'American Clean Energy and Security Act of 2009',[3] as considered in the US-Senate. Furthermore, an independent bill was introduced also seeking to implement a cap-and-trade-system.[4] However, in the aftermath of the oil spill cataclysm in the Gulf of Mexico, the bill has been replaced by a substantially altered draft making the implementation of the intended cap-and-trade system highly unlikely in the near future.[5] The EU Emissions Trading Scheme (ETS), encompassing the energy industry itself and the energy intensive industries (Metcalf and Weisbach, 2009, p. 10), has so far failed to reduce the amount of emitted carbon dioxide in the EU countries; whether the amendment to and the changes of the applicable Directives, enlarging the scope of the concerned industries, will lead to a better success is at present hard to reasonably estimate (Weber, 2008, p.p 484–9).

The key feature of the cap-and-trade scheme consists in the governmental stipulation of a maximum amount (cap) of carbon dioxide that can be emitted and the details of the trade mechanism (Cockfield, 2011, p. 8).

b) Shortcomings

A major drawback of cap-and-trade regimes is the lack of cost certainty, because the price for the allowances may be volatile as trade with 'climate certificates' is subject to basic market mechanisms (law of supply and demand). The assessment of costs is particularly difficult (Weber, 2008, p. 487, Cockfield, 2011, p. 8) if derivative markets would not take over a part of the future-oriented risks (Weber and Darbellay, 2010, p. 271 *et seq.*). Additionally, if allowances are distributed without an auction for free at the beginning, the risk of a price collapse in the market occurs (Weber, 2008, 487, Avi-Yonah and Uhlmann, 2009, p. 42). In the case of persistent volatility, the (theoretical) option exists to auction additional allowances; this procedure is called 'safety valve'. However, it will then no longer be possible to achieve the previously set goal of a maximum amount of emissions due to the increase of the cap (Avi-Yonah and Uhlmann, 2009, p. 42).

Secondly, a cap-and-trade system does not raise any revenue for the state hence the government has not any additional means to fund alternative energy or to alleviate the adverse effects of climate change (Cockfield, 2011, p. 9).

Thirdly, cap-and-trade systems tend to be highly complex in their implementation as well as in establishing robust trading mechanisms; due to their complexity these systems are also more susceptible to inherent misconceptions (Avi-Yonah and Uhlmann, 2009, p. 38). Moreover, the government has to play a much larger role in contrast to a carbon tax, as a cap-and-trade system requires much more administrative efforts, since emissions and offset allowances have to be created (Milne, 2009, p. 457 *et seq.*)

Fourthly, the enforcement of such systems is not an easy task due to its intricacy. It has to be thoroughly monitored in order to prevent manipulation and exploitation by afflicted market participants generating additional costs (Cockfield, 2011, p. 9).

Finally, as in case of a carbon tax, one of the major issues is the absence of a globally applicable and enforceable approach.

III. DEVELOPMENT OF NEW ENVIRONMENTAL TAXATION STRATEGIES

1. Introduction

As set out, both 'traditional approaches' do not offer a satisfactory solution to the problem of 'carbon leakage'. Looking at the present political situation after the Summit of Cancun, alternatives should rather be based on a framework which entails taxation provisions on domestic carbon emitters as well as on goods or services provided by enterprises operating out of countries not complying with an international agreement than on a cap-and-trade-regime. As such strategies (i) an extension of the OECD Model Tax Convention and (ii) newly developed international tax regimes will be considered.

A principle problem consists in the fact that sovereignty issues play a much more important role in taxation than for example in international trade or international finance. Therefore, whatever new taxation strategies are taken into account and whatever new sovereignty concepts (such as co-operative or multilevel sovereignty) (Weber, 2010) are developed, the territorial application scope of national tax laws should not be overlooked.

2.   Extension of OECD Model Tax Convention

For the purpose of creating an internationally accepted carbon taxation regime the OECD could serve as a suitable panel, not least due to the highly esteemed OECD Model Tax Convention on Income and on Capital.

a)   Carbon tax shaped as income tax

Most tax treaties are concluded bilaterally between states; therefore, in order to enlarge the scope of involved states it could be considered to use the widely acknowledged OECD Model Tax Convention as a medium for a carbon income tax.

One major obstacle of any new tax strategy consists in the fact that the imposition of a carbon tax on imports from non-participating countries might violate the national treatment or most-favoured nation principle as regularly stated in trade agreements. However, trade agreements which feature said principles, such as the General Agreement on Tariffs and Trade (GATT), do not interfere with the sovereign taxation of persons or enterprises, but instead they are applicable to indirect taxes on goods or services (Weber and Kaufmann, 2011). This circumstance makes it advisable to design a carbon tax as an income tax; thereby, the participating countries will be able to avoid violating trade agreements.

Since levying taxes is in principle only possible within the borders of a certain state, the question arises how to tax non-resident carbon dioxide emitters. Generally, an income tax may be imposed on non-residents if they maintain a permanent residence in the tax imposing country, as, for example, stated in Article 5 of the OECD Model Tax Convention. If a company headquartered in a country not participating in an international carbon pricing agreement has a permanent residence in a state, which imposes carbon taxes, the profits earned by this corporation may be subject to carbon taxation.[6]

b) Non-discrimination principle

The OECD Model Tax Convention features at present in the 2008 revised Article 24 a specific provision against discrimination in income taxation on the basis of nationality. This provision cannot be interpreted as a most-favoured nation principle according to the commentary on the OECD Model Tax Convention (OECD, 2010a) since it prohibits discrimination solely based on one specific ground, e.g. nationality. Furthermore, the Convention is based on the principle of reciprocity; therefore, taxation of permanent establishments of a contracting state located in another contracting state must not be more burdensome than permanent establishments of this state are subjected to under the same circumstances in the other state (OECD, 2010a, p. 332). If a carbon tax is introduced only in one country, the taxation of a permanent establishment from another country would be prohibited under Article 24, since establishments of the tax imposing state located in the other state would not be subjected to said taxation.

In addition, the adoption of a limited most-favoured nation principle in the OECD Model Tax Convention should be considered. Such principle could be shaped in the form that if country A comes to a more favourable agreement with country B, the new terms would not be automatically applicable to country C, but rather generate an obligation for country A to enter into new negotiations with country C (Cockfield and Arnold, 2010, p. 151). Consequently, countries relying for their tax regime on the OECD Model Tax Convention by introducing a carbon tax get an opportunity to renegotiate a bilateral agreement.

c) Extension to non-OECD countries

As mentioned, the application of a bilateral or a plurilateral tax agreement on enterprises not being domiciled and not executing any business activities in any of the member states of the treaty causes problems. However, the incorporation of the principles of the OECD Model Tax Convention combined with the 'threat' of an extension to non-participating countries could serve as a considerable moral incentive for non-participating states to introduce a carbon tax.

If companies from non-participating countries do not maintain a permanent residence, the imposition of an income tax is not directly possible. In this case a special carbon withholding tax on the providers of goods and services stemming from a non-participating country would have to be implemented. Such a special withholding tax could be levied (i) in form of a consumption border tax adjustment or (ii) in form of an income border tax adjustment (Cockfield, 2011, p. 16/17).

(i)   The introduction of a consumption border tax adjustment (eventually combined with a system of financial transfers[7]) would to lead to an additional carbon tax being imposed on imports of goods from non-OECD countries or on services provided by firms operating out of such states. Such a tax might avert the risk of 'carbon leakage' as firms with high carbon emissions would be discouraged to relocate their businesses to non-participating countries. This concept would especially have an impact on global competition hence industries from OECD countries would be subject to the same carbon tax while an import tax would be levied on enterprises of non-participating countries (Cockfield, 2011, p. 16).

The implementation of this concept might conflict with international trade agreements, such as the World Trade Organization's (WTO's) General Agreement on Tariffs and Trade (GATT) or General Agreement on Trade in Services (GATS), since a consumption border tax adjustment could result in an infringement of the non-discrimination and national treatment principles.[8] Therefore, it is vital that such a consumption tax is not misused as means to disguise protectionist intentions, but rather compensates marginal costs arising from the production of carbon intense products (Cockfield, 2011, p. 16 et seq.).

(ii) The other proposed concept consists in the implementation of an income border tax adjustment for non-OECD countries under the OECD Model Tax Convention. One reason why such a tax should be taken into consideration is that the taxation of goods and services might possibly lead to severe disputes between countries due to a violation of WTO obligations. Nevertheless, if the tax is formed as an income tax, the above mentioned problem is no longer of importance, since trade agreements do usually not apply to income taxes. Such an income tax could be easily levied on companies from non-OECD countries maintaining permanent business establishments. If firms from non-OECD countries do not possess a place of business in a member state a withholding tax could be collected on goods or services provided by such corporations (Cockfield, 2011, p. 17). Thereby, non-OECD countries would become subject to the same level of taxation as member states.

3. Newly Developed International Taxation Regimes

Taking up the evaluation of the traditional instruments in the light of the above outlined taxation strategies, further carbon leakage prevention regimes could be taken into account.

a) Agreement on a preferred pricing mechanism

A possible approach to address the issue of 'carbon leakage' is the development of an international agreement basing carbon taxation on a preferred pricing mechanism scheme. In a first step, all countries participating in an international agreement would have to come to a common understanding concerning the initial tax rate. Obviously, an inherent problem is embedded in the fact that a mutual consent related to the extent of emissions' reduction cannot easily be reached; the chance of the big global players agreeing on terms of emissions' reduction seems improbable. However, a solution could eventually be found in an agreement on a minimum carbon price, which serves participating countries as minimum level for their national carbon pricing (Cockfield, 2011, p. 15). Initially, the price for carbon should commence at a low rate in order to encourage more countries to take part in the agreement and to allow the producers as well the consumers to slowly adjust to the changed circumstances (Waggoner, 2009, p. 13). Furthermore, the development of alternative energy sources would be given time before the price of carbon has become too expensive (Waggoner, 2009, p. 7).

Since an internationally agreed carbon tax based on a preferred pricing mechanism would presumably lead to a predictable price on carbon dioxide and be comparatively easy to enforce,[9] it may be appealing to states to implement such kind of tax to the extent that national policies are complying with international standards (Cockfield, 2011, p. 15). If a government is participating in a such an agreement and actively addresses climate change it should have a signalling effect on the industries of said country to invest in 'greener technologies' in order to save expenses (Cockfield, 2011, p. 15 *et seq.*). Consequently, through the thereby triggered innovation these producers will be provided with a competitive advantage comparing to high carbon emitters in low standard states, which will create an incentive for such states to introduce a carbon tax in order to ensure the competitiveness of their industry (OECD, 2010, p. 125 *et seq.*).

b) Carbon sequestration credit system

For the improvement of the efficiency of new taxation strategies it is important that not only a carbon tax is imposed on the emitters, but also that the permanent sequestration of carbon dioxide is awarded. For example, the technology referred to as carbon capture and storage (CCS) is one possibility to safely capture carbon and store it permanently underground (Metcalf and Weisbach, 2009, p. 37 *et seq.*). Thus, the implementation of a carbon sequestration credit system could be taken into consideration as means to create a powerful incentive to capture and store carbon dioxide emissions or to develop sophisticated technologies which enable such undertakings (Metcalf and Weisbach, 2009, p. 38).

This concept, however, will only be truly effective if the use of carbon producing materials is excluded from the tax, provided that carbon dioxide arising from this process is subject to sequestration. Additionally, credits distributed due to the capture and storage of emissions should be tradable, since the producer of the emissions and the provider of the sequestration may not be identical; thereby, the value of these measures can be fully realized by firms conducting carbon sequestration (Metcalf and Weisbach, 2009, p. 38).

c) Hybrid system

Some further, as yet not extensively elaborated, approaches exist. For example, a proposal has been put forward which tries to amalgamate the tax-and-cap concepts into a hybrid tax-and-cap system (Nordhaus, 2008, p. 162 et seq). Following this approach, a tax and a quantitative restraint could be implemented in the same system. Subsequently, a safety valve would have to be established, wherein allowances exceeding the cap could be offered at a multiple of the tax (Nordhaus, 2008, p. 163 *et seq.*). Such hybrid system could gain efficiency and flexibility if the corresponding instruments would be introduced in financial markets through financial intermediation (Weber and Darbellay, 2010, p. 293 *et seq.*).

Through this approach some advantages of both main concepts could be combined. Having the benefit of a quantitative limit of carbon dioxide emissions, a hybrid system would raise revenue and could mitigate cost uncertainty, since the price on carbon is set (Nordhaus, 2008, p. 164). However, irrespective of which approach is ultimately favoured, the issue of how to induce non-participating countries to join such an agreement remains a critical issue.

d) Carbon tax as consumption tax combined with a system of financial transfers

Another possibility to reduce the general carbon dioxide output is not to tax the producers of carbon dioxide, but rather impose a tax on the consumption of carbon dioxide relevant goods or services on the national level. Such a carbon tax on goods and services, designed similarly to a consumption tax, would most likely share many common traits with a regular consumption tax, since carbon dioxide plays a part in the production, distribution or provision of countless goods or services, for example, virtually all synthetic materials consist of carbon in one way or another (Waggoner, 2009, p. 14).

Generally, a consumption tax is regressive in nature, meaning that the tax collects a higher percentage of low incomes then it does on high incomes, being exposed to the situation that persons with low income have to spend almost all their income on consumption, since they have to invest the majority of their resources in food, clothing, rent for lodging and other necessary expenses (Waggoner, 2009, p. 14 *et seq.*). In contrast, people with high income have the possibility to save or invest parts of their resources, thus decreasing the share of their income which is consumed by a consumption or carbon tax (Waggoner, 2009, p. 15).

For social reasons, it is desirable to try to reduce regressivity in order to relieve people of low income from this tax burden. For this purpose, the implementation of a system of financial transfers should be considered. For instance, a tax credit system could be introduced, wherein people with low income would receive a refundable tax credit, which would render the carbon tax less regressive (Waggoner, 2009, p. 16). However, the question remains, whether such a concept would still provide sufficient incentives for people with low income to reduce their purchase of highly intense carbon goods. Since the amount of income would be the basis of assessment for the tax credit, the more people entitled to said refunds would refrain from purchasing carbon intense products the more they could keep the tax refunds. Thus, this concept creates a financial incentive for people with low income to reduce carbon consumption, thereby serving the goal of carbon dioxide emissions' reduction (Waggoner, 2009, p. 17).

4. Challenges During the Present Interim Period

A rectification and enlargement of the OECD Model Taxation Convention will not be realized within the near future. The discussed alternatives for taxation strategies are also unlikely to be implemented in the short term. During the interim period, most likely only an improved unilateral border tax regime is feasible.

a) Unilateral border tax regime

Already intensively discussed is the attempt to incentivize countries with inferior standards to comply with stringent environmental provisions by the imposition of a unilateral tax on imported foreign goods, being associated with high carbon dioxide emissions (Weber and Kaufmann, 2011).

This concept can either be shaped as a carbon tax or as an obligation to purchase emission allowances. Such an approach would unfold its effect the more efficiently the more states would be willing to participate in an international agreement, which stipulates a mutually agreed carbon price or, in case of a cap-and-trade system, a maximum amount of emissions (cap).

Not only would the introduction of a unilateral border tax establish an incentive for non-participating countries, but such approach would also discourage industries to relocate to countries with insufficient environmental standards as their products would again be subject to substantial taxation in case of cross-border trade. Moreover, distortion of competition due to diverging environmental provisions could be mitigated, because all participating countries would be subject to the same carbon tariff, whereas enterprises operating out of non-participating states would have to deal with carbon taxation in a participating country (Cockfield, 2010, p. 16). However, the unilateral border tax would have to be shaped in a way that it complies with international trade agreements, in particular the GATT.[10]

b) Problem of compatibility with WTO law

With the taxation of countries not participating in an international agreement, the issue arises whether such tax could be qualified as a violation of WTO law, in particular the GATT or the GATS. Any unilaterally introduced border tax adjustment (BTA) measures risk to clash with the national treatment and the most-favoured-nation principle. A detailed assessment of this issue is beyond the scope of this chapter (Weber and Kaufmann, 2011); however, a short summary of the issues will be given.

A taxation of goods of a non-participating country is only legal if like products in the state imposing said tax are subject to the same taxation. Generally speaking, carbon taxes on imports can be qualified as indirect product taxes as long as there is a 'nexus' between the tax and the product. Such a nexus exists when carbon taxes aim at creating a level playing field between like products in the country of destination. Therefore, the question has to be answered, whether 'likeness' of two products is given. In principle, the criterion of 'likeness' does not include the process of how a product is manufactured, but rather physical characteristics, consumers taste and habits, the products end use in a given market and the products tariff classification (Weber and Kaufmann, 2011). Nonetheless, the question remains whether products manufactured with high carbon dioxide outputs are truly the same as such which did not include the emission of environmentally harmful gases. The majority of the legal doctrine tends to assume that (i) energy or pollution taxes are to be qualified as indirect taxes; (ii) taxes applied on inputs, such as energy used during the production process, are product taxes; (iii) BTA is possible on a final product for energy and pollution taxes.

Since the compatibility of carbon-related BTA measures with WTO law is still partly contested, potential justifications become relevant. Article XX GATT justifies a violation of the GATT based on legitimate non-trade policy goals provided that such interests are adequately balanced against the objective of free trade. (i) A violation of the GATT can be justified if the measure taken fits under the heading of any specific exception clause in Article XX (a)–(j) GATT, in particular relating to the conservation of exhaustible natural resources or the protection of human, animal or plant health. (ii) If a measure falls under any of these provisions, it still has to be necessary and proportional (Weber and Kaufmann, 2011). However, environmental or health policy choices made by governments gain importance in the dispute resolution proceedings, if the focus can be laid on the relationship between the measure at stake and the legitimate environmental policy.[11] (iii) Furthermore, the BTA measures must be indispensible to reach the national policy goals, i.e., no other less restrictive means should be readily available. Additionally, any kind of taxation will not be justified if it is to be considered as a means to disguise trade restrictions (Metcalf and Weisbach, 2009, p. 49).

IV. INTERNATIONAL FEASIBILITY OF NEW ENVIRONMENTAL TAXATION STRATEGIES

An international agreement on carbon taxation or on a cap-and-trade system will be very difficult to realize unless the two largest carbon dioxide emitters of the world, namely China and the United States,[12] agree to take part in such an endeavour. The political situation does not look very promising for the time being; nevertheless, China could consider entering into such an agreement, as it mainly exports goods to the United States that are not associated with high carbon dioxide emissions (Cockfield, 2011, p. 18), and the United States might be receptive for an international agreement, since the government could address its trade imbalance with certain states, in particular China.

Notwithstanding this appreciation, it remains doubtful that in case of non-participation of China and the United States an international agreement would be able to accomplish the set goal of coming to a halt or at least a mitigation of climate change through substantial carbon dioxide emissions reduction, since these two countries are by far the biggest carbon dioxide emitters in the world and a carbon withholding tax levied on those countries would hardly be politically feasible due to their economic importance.

Even if an international consensus could be achieved, the problem remains that less developed countries would be subject to the same tax rate or cap as wealthy states. This fact might discourage developing or emerging countries to join an international agreement even tough their export industries would be heavily afflicted by a carbon tax. A possible way to mitigate this problem is to implement a tax rate which differs between rich and poor countries. An alternative possibility, in case of a uniform carbon tax, consists in the creation of a distribution scheme which would foresee that wealthy nations provide poor countries with some of their carbon tax income in exchange for their accession to an international agreement, thereby avoiding that the benefit of a carbon tax would be compromised (Sandmo, 2009, p. 22 *et seq.*).

NOTES

1. www.financial-dictionary.thefreedictionary.com/Direct+taxes, accessed 26 November 2010.
2. www.financial-dictionary.thefreedictionary.com/indirect+tax, accessed 26 November 2010.
3. 111th US Congress, H.R. 2454, 6 July 2009.
4. Clean Energy Jobs and American Power Act, 111th US Congress, S. 1733, 30 September 2009.
5. Clean Energy Jobs and Oil Company Accountability Act of 2010, 111th US Congress, S.3663, 28 July 2010; ictsd.org/i/news/bridgesweekly/81833/, accessed 26 November 2010; www.opencongress.org/articles/view/2072-The-Inside-Story-of-How-the-Climate-Bill-Died, accessed 26 November 2010.
6. See below III.2.c.
7. See below III.3.d.
8. This issue will be discussed below under III.4.
9. See above II.1.a.
10. See below III.5.
11. A country needs to demonstrate that the measure is apt to produce a substantial contribution to the achievement of its objective.
12. www.ucsusa.org/global_warming/science_and_impacts/science/graph-showing-each-countrys.html, accessed 26 November 2010.

REFERENCES

Avi-Yonah, R.S. and D.M. Uhlmann (2009), 'Combating Global Climate Change: Why a Carbon Tax is a Better Response to Global Warming than Cap and Trade', *Stanford Environmental Law Journal*; **28**(3); *University of Michigan Public Law Working Paper No. 117.*

Cockfield, A.J. (forthcoming 2011), 'Optimal Global Warming Tax Policy for Small Open Economies', in Richard Cullen and Jefferson D. Vanderwolk (eds.), *Green Taxation in East Asia*, http://papers.ssrn.com/sol3/papers.cfm?abstract_id= 1596085, accessed 26 November 2010.

Cockfield, A.J. and B.J. Arnold (2010), 'What Can Trade Teach Tax? Examining Reform Options for Art. 24 (Non-Discrimination) of the OECD Model', *World Tax Journal*, 2/2010, 139–53.

Duff, D.G. (2003), 'Tax Policy and Global Warming', *Canadian Tax Journal*, **51**(6), 2063–118.

Enders, A. and C. Ohl (2005), 'Kyoto, Europe? – An Economic Evaluation of the European Emission Trading Directive', *European Journal of Law and Economics*, **19**(1), 17–39.

Metcalf, Gilbert and David Weisbach (2009), 'The Design of a Carbon Tax', *University of Chicago Law School John M. Olin Law & Economics Working Paper No. 447/2009*.

Milne, Janet E. (2009), 'Carbon Taxes Versus Cap-and-Trade', in Ling-Heng Lye, Janet E. Mils, Hope Ashiabor *et al.* (eds.), *Critical Issues in Environmental Taxation*, Oxford: Oxford University Press, 445–62.

Nordhaus, William (2008), *A Question of Balance*, New Haven and London: Yale University Press.

OECD (1980), *Pollution Charges in Practice*, reprinted in Kalle Määttä (2006), *Environmental Taxes: An Introductory Analysis*, Cheltenham, UK and Northampton, MA, US: Edward Elgar Publishing.

OECD (2010), *Taxation, Innovation and the Environment*, Paris: OECD.

OECD (2010a*), Commentary on OECD Model Tax Treaty*, http://browse.oecdbookshop.org/oecd/pdfs/browseit/2310081E.PDF, 332, accessed 26 November 2010.

Pigou, Arthur C. (1932), *The Economic of Welfare*, London: Macmillan and Co.

Sandmo Agnar (2009), *The Scale and Scope of Environmental Taxation*, Norwegian School of Economics and Business Administration Discussion Paper No. 18/2009, http://ssrn.com/abstract=1554948, 14, accessed 26 November 2010.

Stern, Nicholas (2007), *Report on the Economics of Climate Change, The Stern Review*, Cambridge: Cambridge University Press.

Waggoner, M. (2009), 'How and Why to Tax Carbon', *Colorado Journal of International Environmental Law and Policy*, Vol. 20; U of Colorado Law Legal Studies Research Paper No. 09–06.

Weber, Rolf H. (2008), 'Emissions Trading', in Nedim P. Vogt, Eric Stupp und Dieter Dubs (eds.), *Unternehmen – Transaktion – Recht, Liber Amicorum für Rolf Watter zum 50. Geburtstag*, Zurich: Dike, 475–92.

Weber, R.H. (2010), 'New Sovereignty Concepts in the Age of Internet?', *Journal of Internet Law*, **14**, 12–20.

Weber, R.H. and A Darbellay (2010), 'Regulation and Financial Intermediation in the Kyoto Protocol's Clean Development Mechanism', *The Georgetown International Environmental Law Review*, **22**(2), 271–360.

Weber, R.H. and C. Kaufmann (forthcoming 2011), 'Border Tax Adjustment in View of Climate Improvement', *World Trade Review*, **10**.

World Bank Group (1998), 'Pollution Charges: Lessons from Implementation', *Pollution Prevention and Abatement Handbook*, Washington D.C.: The International Bank for Reconstruction and Development/THE WORLD BANK, 160–8, www.ertc.deqp.go.th/ertc/images/stories/user/ct/ct1/cp/cp_

program_management/WB%20on%20Pollution%20Charges.pdf>, accessed 26 November 2010.

WTO (2009), *Trade and Climate Change*, WTO-UNEP Report. Geneva: World Trade Organization.

WTO (2009a), 'Background Note: Trade and Environment in the WTO', www.wto.org/english/news_e/news09_e/climate_21dec09_e.pdf, accessed 26 November 2010.

Xu, Yan (2010), 'Green Taxation in China: The Case for Consolidated Fuel Transportation', 2nd Taxation Law Research Program (TLPR) International Conference on Green Taxation in East Asia: Problems and Prospects, Hong Kong, 29 January 2010, www.hku.hk/law/research/AnnualReport.pdf, accessed 26 November 2010.

Yandle, B. (1999), 'Public Choice at the Intersection of Environmental Law and Economics', *European Journal of Law and Economics*, **8**(1), 5–27.

5. Why should there always be a loser in environmental taxation?

Sally-Ann Joseph

A. INTRODUCTION

Despite the lack of an unequivocal conceptual definition of resilience and its relationship to other key concepts such as vulnerability, sensitivity and adaptive capacity, the economic dimension to ecological degradation is clear: environmental damage, biodiversity loss and depletion of natural resources reduce corporate productivity (Salzmann, Ionescu-Somers and Steger, 2005). Reduced corporate productivity negatively impacts on corporate profitability resulting in decreased taxation revenues. Climate change resilience requires the implementation of measures that proactively increases productivity by reducing climate impacts.

To achieve this either there must be a drastic reduction in consumption of resources or there must be a drastic increase in resource productivity. The latter is more attractive, economically, socially and politically.

Two major environmental or climate change concerns are associated with the business sector. These are (1) the use, and subsequent depletion, of non-renewable resources and (2) pollution or the contribution made to global warming through emissions. To have an impact on these requires large-scale change in corporate behavior. But equally, the current framework for doing business does not appear to reward those companies trying to be sustainable. Measures taken to date have not aligned corporate and environmental goals resulting in there always being losers (actual or perceived) requiring some form of compensation by the government. Whether monetary recompense or exemptions from the regime, the overall complexity of the system is increased and the benefits watered down thus extending the concept of losers beyond those originally impacted even as far as to the broader community.

Part B looks at the instruments available to influence corporate decision-making, highlighting their shortcomings. Part C then analyzes selected approaches to environmental taxation to determine how effectual or ineffectual they have been in changing corporate behavior. What drives corporate decision-making with respect to sustainability is canvassed in Part D, followed by a conclusion in Part E.

B. AVAILABLE MEASURES

The measures available to effect environmental taxation can be classified as economic instruments (market-based and financial), command-control measures and voluntary measures.

Economic Instruments

Economic instruments are increasingly being used for the management of resources. They are not 'just another tax'. Indeed, in some cases they involve no taxation at all. Their purpose may be to change the relative prices of goods and services and thereby to change behavior, not necessarily to raise revenue. All economic instruments, both market-based and financial, rely on a legislative framework to establish their validity and to ensure enforcement.

The most important economic instruments in relation to environmental management are tradeable permits, environmental taxes and special charges. Others include deposit refund schemes, performance bonds and subsidies.

Tradeable permits are the latest fashionable instruments for achieving environmental goals. They are structured as incentive-based as the instrument should contribute to attainting predetermined environmental thresholds, goals or standards. Permissible consumption and emission levels are set for which permits are issued for their total volume. What is considered a 'permissible' level will often be based on a political compromise.

Such a system, however, presumes an allocation of permits and some method of measuring emissions to determine compliance. Allocation is a preeminently political decision – it is an explicit 'right to pollute'. It requires the government to delineate property rights over the use of resources then relies on the market to determine the distribution of those rights. Monitoring is technically contentious and presupposes continuous, accurate and credible emission measurements. A cap-and-trade system affords flexibility, the 'price' of which is that any single company may choose not to reduce emissions at all. So, although designed as incentive-based, they do not provide corporate entities with a predominant motivation to be environmentally sustainable.

Special charges artificially raise the price of the resource, pollution or other environmental concern. They generally have a financing function with the proceeds earmarked for a particular use such as services, remedial action or prevention pertaining to that concern. They may also serve to influence behavior.

Tax earmarking requires proof of the causal link between the taxed commodity and the environmental damage. Thus earmarking requires hypothecating. Because of this burden of proof, earmarked charges are subject to relatively narrow limits. Special charges can be referred to as fees, licenses, levies, excises, royalties, or charges, amongst other names. In practice politicians decide what label to give a new impost.

By fixing the use of tax revenue in advance, earmarking creates inflexibility. Programs may last longer than is optimal as vested interests obstruct reform even when policy priorities have changed. Earmarking also frequently results in inefficient spending of government revenue. For example, allocating transport taxes to road infrastructure may lead to over-investment in that sector.

Special charges are generally price instruments meaning that the marginal cost or price is fixed but not the quantity. They come in many forms and disguises and may be direct or indirect. User charges, for example, are levied on users of publicly supplied services, the revenue of which helps to defray the cost of providing the service. They can be restricted to paying consumers such as entry fees to national parks. Or it may be too costly or impractical to charge directly for the service and imposed on some good needed to use the service. A road user charge can be imposed as an annual license fee for a motor vehicle or as a charge on fuel (or both).

Environmental taxation is a special charge. Carbon taxes have been widely advocated as they can be imposed precisely on the source of the environmental problem. Indeed, they are said to be preferable to an emissions trading scheme from an economic perspective (Hepburn, 2010; Avi-Yonah and Uhlmann, 2009). Yet their implementation is the exception rather than the rule (OECD, 2002; Byers, 2010).

One of the more commonly imposed environmental taxes is effluent charges. The amount payable under these license fees can be adjusted according to the size of environmental harm or risk of the activity. This would appear to provide the licensee with an incentive to reduce their environmental footprint. However, the relatively small size of the fee and the discretionary way in which it is imposed is insufficient to change practices. In addition, they are viewed more as a 'tax grab' or 'revenue raiser' than mitigating environmental harm. This is largely due to the fact that the factors used in assessing risk, such as compliance with industry codes of practice and evidence of contingency plans, are not tied to the environmental impact. As a result, these generally reflect only the physical costs of waste disposal rather than the cost to the environment. This is a common issue with most, if not all, environmental taxes and special charges. The level to which the appropriate tax should be set in order to be effective may be so high as to be impracticable.

Even where there are penalties and liability for non-compliance, affected companies may opt to pay the penalty rather than incur the higher marginal cost of meeting the cap or ensuring compliance. Here the environment is the loser. Where the costs result in higher prices, consumers are the losers. Governments may act to mitigate this in a number of ways. They may decide to compensate consumers, or a segment of consumers such as pensioners, either by direct monetary recompense or tax credits. Alternatively, or in conjunction, governments may grant certain industry sectors exemptions or relief. Governments of Organisation for Economic Co-operation and Development (OECD) countries currently provide over 1150 exemptions to environment taxes and several hundred refund mechanisms and other tax provisions (OECD, 2007). Both reduce revenue making government the loser. Both create equity issues making society the loser. But business is seldom, if ever, the loser.

Command-control Regulatory Measures

Command and control measures, also referred to as standards or regulations, directly prohibit or restrict activities that harm the environment. The traditional process consists of legislating on regulatory measures and delegating their implementation and administration to a regulatory authority. The 'command' sets a standard (such as the maximum level of permissible pollution) and the 'control' monitors and enforces the standard. Thus, this approach generally relies on detailed regulations followed up by an ongoing inspection program.

Regulation forces immediate compliance regardless of the comparative costs for different businesses within and between different industries. In addition, there is an implicit assumption in prescribed standards that cost functions are identical across sources (Ellerman, 2000). Command-control measures do not allow firms to find other cheaper ways to achieve the same result. The financial costs may be high. There could also be political costs if the measures are stringent and businesses are adversely affected. To be effective, regulatory measures need to be revised frequently to keep up with changes and economic developments. This entails legislative amendments which tend to lag practical developments.

An 'optimum' standard is difficult to determine, especially with non-marketable goods such as water and air. And then there is no incentive for businesses to reduce pollution or the targeted environmental harm beyond the standard. Penalties for violating standards tend to be too low and enforcement weak. For the sanction to be effective, the cost of non-compliance must exceed the benefits of non-compliance. Sanctions may be correlated to the environmental damage. But this implies that the environmental damage can be measured or quantitatively ascertained in some way. And then the regulator will be obliged to accurately account for the specific contribution of each transgressor. That requires vast information which is not only asymmetric and therefore difficult to obtain, but also costly to produce.

Regulation, with its command-control measures, is supposedly the response to public goods and free-rider problems, collective action and information failures, tragedies of the commons, and negative externalities (Ahdieh, 2009). It is evident that these issues continue unabated. Unlike a price instrument that fixes the marginal cost or price but not the quantity, or a quantity instrument that fixes the quantity but not the price, a regulatory standard does neither. They are purely punitive measures – all stick and no carrot.

Everyone is a loser here. For businesses it is the lack of flexibility and increased costs (and therefore decreased profits). For regulators it is administrative costs and sub-optimal revenue stream associated with its difficulty to police and maintain currency in light of developments in business practices and environmental issues. The environment loses, firstly, because there is no incentive to do more than the absolute minimum and, secondly, because the system is not as effective as it could be. Unresolved matters such as free-rider problems and tragedies of the common also impose societal costs, making society losers.

Voluntary Measures

Voluntary measures is a broad term that encompasses many different kinds of arrangements, such as self-regulation, voluntary initiatives and codes, environmental charters, voluntary agreements and negotiated agreements. They are commonly classified into three main types: unilateral commitments, public voluntary programs and negotiated agreements (Couder and Verbruggen, 2005).

Unilateral commitments are environmental improvement programs initiated and implemented by business. Their content, such as targets and time schedules, are decided upon by the businesses and/or industry associations involved in the program. The role of stakeholders (shareholders, employees, consumers) is merely as recipient of communications. The driver for business is generally image and perception in the market. It is being seen to do 'the right thing'. However, targets are not usually over-ambitious and commitments can be restricted to those areas where costs can be controlled.

Public voluntary programs are schemes initiated by government, usually public authorities. Here businesses can voluntarily join by agreeing to meet the targets and/or adhere to the standards set by the public authority. A motivating factor for businesses is conformity to industry practices. This may be a result of collective/group behaviors or peer pressure. Doing something for the wrong reason is itself unsustainable.

Negotiated agreements are similar to public voluntary programs in that they are contracts between public authorities and businesses (including with industry associations). The difference is that the targets (such as time schedules or quantitative measures) and/or standards (such as processes or procedures) are negotiated between the parties. In return, the public authority generally refrains from introducing more coercive or costly regulatory measures. The 'incentive' is punitive.

These arrangements are essentially designed to change the behavior of corporations without direct government intervention. They rely more on education and persuasion than on incentives and power imposition (Burritt, Lewis and James, 2005). Participation sends signals to stakeholders who modify their behavior, favorably, towards the committed business. This may result in increased market share, being an employer of choice or reflected in the share price. In this way a benefit is created which provides the participating corporation with a competitive advantage over its uncommitted competitors. However, participation is not mandatory. Non-compliance is punished only through the loss of the advantages linked to the signals although there may also be moral sanctions.

An OECD report concluded that 'the environmental effectiveness of voluntary approaches is still questionable', and 'the economic efficiency of voluntary approaches is generally low' (OECD, 2003, p.14). It also found that there were only a few instances where voluntary approaches contributed to environmental improvements significantly different from what would have happened anyway under 'business as usual'.

If there are any winners it is because of business self-interest. Otherwise everyone, and everything, loses.

C. CURRENT APPROACHES TO ENVIRONMENTAL TAXATION

Three of the more dominant approaches to environmental taxation are (1) internalizing externalities using the polluter-pays principle, (2) using technology as a means of providing solutions and (3) large-scale reform commonly referred to as environmental or ecological tax reform.

The Polluter-pays Principle

The polluter-pays principle is popular with both policymakers and the public. It appeals to the basic principles of fairness and justice: people and entities should be held responsible for their actions. Those who cause damage or harm should pay accordingly. Forcing entities to bear the costs of their activities is also said to enhance economic efficiency (Mann, 2009).

The polluter-pays principle is the chief economic pillar of a market-oriented environmental policy. Its objective is to shift the responsibility of dealing with pollution, waste or other environmental damage from governments to the entities producing or causing it. Application of this principle has spawned a myriad of command-control measures. These may take the form of special charges or specific standards and/or regulations. More recently the policy has shifted to a market-based approach. Whether taxes or tradeable permits, economic instruments are seen as attempts to make the polluter pay by attaching a price to the polluting activities that were previously undertaken for free. In the case of taxes, this 'price' is determined by the government; for tradeable permits the 'price' is determined by market supply and demand conditions.

However, there are practical implications in applying the principle. What happens when repair is impossible (such as in the case of species extinction) or the damage is impossible to quantify (such as the effect of greenhouse gases)? Rarely can monetary compensation make up for biological loss or loss of a resource such as artesian water or topsoil. In reality, to some degree at least, the polluter can never pay the real cost of their impact even when some restoration is possible.

What happens if the polluting company no longer exists or cannot be categorically identified? Assigning or apportioning responsibility is fraught with difficulties, practical, legal, even political. In many instances causal chains are not only hypothetical but also controversial. Thus environmental damage still occurs for which either no one pays or which is paid for by all citizens – the common burden principle.

And then there is the question of what instrument to use. When targets are determined arbitrarily without knowing the costs of production, via market signals or otherwise, no instrument is efficient. The primary objective has developed so that it is no longer necessarily environmental but rather to remedy distortions in international trade and investment. This development resulted from the employment of different pollution abatement financing methods which adversely impacted on the costs of production and international competitiveness of companies domiciled in countries that had adopted the polluter-pays principle (Grimeaud, 2000). Further, pollution charges have often been viewed as having been established primarily as revenue sources rather than to further environmental aims (Laves and Malès, 1989). Finally, internalizing costs becomes a cost of doing business. As such, this cost is passed to customers in the price of the good.

The polluter-pays principle has not been effectual in changing corporate behavior. It is vague and uncertain in its application: can it apply? Even, will it apply? Totally or partially? It is perceived as contradictory with hidden agendas; it has no teeth. In the end it is society that pays, through higher prices, depleted environment, diversion of public funding and increased skepticism in the government to name just a few.

Technological Development

Technological advances have made possible new sources of energy, new materials, increased industrial productivity, global communications and remediation of environmental damage. For example, the Danish tax on the sulphur content of energy products has had a positive impact on the development of sulphur purification plants and technology (Laskowska and Scrimgeour, 2002). Technology has also enabled information and transaction costs to fall substantially, making feasible an ever-increasing range of approaches.

Traditionally technological development has been generally limited to incremental improvements to existing technologies. Even what is considered to be 'innovation practices' is incapable of delivering technologies and business plans compatible with sustainability (Weaver, Jansen, van Grootveld *et al.*, 2000). Indeed, technological innovation has focused on reducing the use of labor, which is taxed, instead of natural resource use which is effectively susidised. In addition financial criteria and short-termism dominates. Greater cognisance is taken of payback periods and unknown technologies intrinsically increase risk. Economic instruments that would stimulate investment in new environment-related technologies have been insufficiently used (OECD, 2002). An example is given of carbon taxes that could provide economic incentives for research on climate change technologies had they been implemented at appropriate levels or rates or on a sufficiently broad scale.

For modeling long-term environmental problems such as climate change, the effects of technological change compounded over long-term horizons are likely to be large. Assumptions are necessarily made about the future rate and direction of technological change. With long lead times and long development times, it is risky determining what technologies will be, firstly, required and, secondly, most effective (OECD, 2002). This is the province of research and development (R&D).

However, a company that invests in a new technology typically creates benefits for others while incurring all the costs. Just as pollution creates a negative externality (which results in the market allowing too much of it), so technology creates positive externalities (so that too little of it is produced). Hence, the market doubly underprovides for new technology for pollution reduction. In other words, where the benefits are more public than private, there is a tendency towards inadequate technological innovation.

As free-market systems are likely to under-invest in R&D, there is a role here for government. However, direct public R&D may crowd out private research efforts or provide windfall gains to those who would have undertaken the research even in the absence of government involvement. Of more concern is the potential for government to pick winners – subsidising the development of particular technologies, techniques or approaches.

An aspect of 'picking winners' can be found in command-control regulations that force the development and/or diffusion of particular technologies. This creates technology 'lock-in'. But this is not restricted to command-control initiated technologies. If the government encourages a particular technology, the risk is that it could become so entrenched in the market place that it prevents, at least for a time, the development of some other superior technology. Thus, old technology may not only be locked in but may also lock out new technology.

Command-control regulations also typically spawn end-of-pipe technologies. End-of-pipe technology is an approach to pollution control that concentrates on treatment of waste prior to discharge into the environment as opposed to making changes in the processes giving rise to the waste. Thus it is an attempt to 'cure' existing environmental issues (such as with the use of pollution filters) rather than to seek innovative and proactive solutions and preventative methods (Georg, 1994).

Technology is capable of driving or supporting sustainability but it will not happen automatically. It will require considerable effort to influence long-term research, technological development and innovation practices geared towards sustainability. The shift needed is akin to another industrial or information revolution.

Environmental Tax Reform

Many countries, particularly in Europe, have implemented large-scale environmental or ecological tax reforms (ETRs). These have been premised on (1) being 'revenue-neutral' and (2) providing a 'double-dividend'. Further, unlike earmarked taxes, no proof of the causal link between the taxed commodity and the environmental damage is required.

Being revenue-neutral entails shifting the burden of taxes from conventional taxes to environmentally related activities so that there is no net increase in the total tax-take. Switching from labor or capital taxation to environmental taxation necessarily narrows the taxation base from which revenue can be raised. There is nevertheless the expectation that, as the tax on (say) energy increases, the tax on (say) labor should decrease. This is not the case – the two trends are parallel rather than opposite (Eurostat, 2008). The shares of revenue raised by labor, capital and environmental sources have remained constant (Pezzey and Park, 1998) and therefore there has been no shift in real terms. This may be due, in part, to a low rate of energy tax coming off a narrow tax base which is deemed necessary as employment may only increase if the environmental tax rates are low (Bovenberg and van der Ploeg, 1998).

Further, as revenues generated from environmental taxes are only a small fraction of the tax, or more commonly, national insurance/social security contributions they replace, it cannot be expected that an ETR will lead to dramatic reductions in these contribution rates. For example, the United Kingdom's climate change levy was introduced on a revenue-neutral basis in 2001. Employer contributions to national insurance (a non-wage labor tax) were reduced by 0.3 per cent, from 12.2 per cent to 11.9 per cent (Williams, 2008). However, they were increased to 12.8 per cent in 2003 (HM Revenue and Customs). This experience is common to all countries that implemented an ETR on the back of reduced social security/national insurance contributions. It could be argued that the rate would have now been higher in the absence of an ETR. Or that rates have been necessarily increased due to other policy factors. These are difficult to contest. However, the link between the environmental tax and reduced national insurance/social security burden has disappeared as far as taxpayers are concerned, leaving the perception of failure and broken promises.

While introduction of ETR may have been revenue-neutral overall, its impact on individual businesses varied substantially. Reducing labor-related taxes benefit labor-intensive businesses disproportionally. It is also worth noting that the rate applied to energy-intensive industries is usually considerably less than that applied to other energy users. This seriously undermines the effectiveness of carbon/energy taxes. It also increases the social cost of achieving an environmental target.

The notion of a double-dividend provides for two traditional benefits: an environmental benefit being ecological sustainability and an economic benefit of low unemployment. The overall consensus is that ETR is unable to deliver the double-dividend (Lawn, 2006).

The environmental effects have been illusive. Scandinavian countries were among the first to introduce environmental tax reform. Yet only Denmark has seen a large decline in carbon emissions while Norway's emissions increased by over 10 per cent between 1990 and 2002 (Lawn, 2006; Prasad, 2009). Suboptimal results appear common. Further, it appears that carbon dioxide emissions are lower only in the short term (Lawn, 2006).

The effect on employment is inconclusive. A review of the German ETR revealed not one single business representative interviewee noting any employment effects since its inception in 1999 (Beuermann and Santarius, 2006). In fact, the policy-makers, business representatives and general public focus group participants almost unanimously supposed that the ETR has neither environmental nor employment effects (Beuermann and Santarius, 2006). According to the French review interviewees, the tax design was evidence of the complete lack of environmental goals – the government merely needed fiscal revenues to fund the '35-hours' employment program (Deroubaix and Lévèque, 2002). Again this illustrates the importance of perception for acceptance.

The German review also notes that having a high level of environmental awareness is inconsequential. It does not translate to a willingness to adopt national policies and measures. Nor, apart from one company, did the researchers find much effort of the business representatives to implement strategies beyond improving the cost-effectiveness of their respective companies (Beuermann and Santarius, 2006). Indeed, the French review stated that the business consensus was that, if some energy savings were possible (which was actually doubted by some interviewees), the tax had nothing to do with it. It would be solely contributable to the profit-maximisation objective (Deroubaix and Lévèque, 2002).

The ETR was widely accepted to be a policy that simultaneously advanced both economic and environmental goals. This appears not to have eventuated, at least to the degree anticipated. It will be necessary to determine if this can be attributed to design and implementation or if the theoretical principles cannot translate to practical application.

D. SUSTAINABILITY DRIVERS TO CORPORATE DECISION-MAKING

There is considerable literature on how corporate entities can become more sustainable. There is also much literature on the positive correlation between environmental and financial performances (Murphy, 2002). If the steps have been formulated and the benefits articulated, why are not all companies further advanced along the environmental sustainability path?

Not much is currently known about the corporate decision-making process with respect to sustainability generally, and environmental issues in particular. Nevertheless, there are two high-level issues that, if aligned, should speed up the process. These are (internal) risk and (external) design.

Risk

For a corporate decision-maker, uncertainty equals risk, especially uncertainty where the outcome is not determined temporally or spatially. Climate change is this kind of risk. However, an unknown outcome is not novel – business can learn from the Y2K experience.[1] That there were no major incidents is either a vindication of the preparation process or the problem was severely overstated.

Corporate decision-makers are concerned with the effect an action taken in the present has upon the options available in the future. Yet it is important to qualify and understand 'risk'. Distinguishing between short-term profits and long-term sustained prosperity is not 'risk'; it is strategy. Likewise is integrating stakeholders and/or their concerns into the decision-making process thereby reducing the pressures exerted or re-thinking potentially defunct products and processes that can create opportunities for improvements and innovation.

The primary opportunities to pursue the twin goals of economic development and environmental quality are technological in nature. Collaboration can mitigate the inherent risks. Practically the government cannot subsidize all new technologies so there is a need to focus scarce resources on commercialisation opportunities for which there is the clearest need for a public role. The private sector may have better information than the government about the likely commercial feasibility of technologies and hence be more successful at identifying which technologies to pursue. Sustainable technologies need to be viable not just technically feasible. While a lack of research makes it difficult to draw conclusions, it appears that the successful programs (those that were actually or potentially of real commercial value) are ones which involved significant participation by industry (National Research Council, 2001).

What the government can and should do is to provide an innovative climate. Technological development cannot happen in isolation. It requires the necessary structural and cultural conditions for successful technology diffusion. Where technology has had its most advancement is where there has been consensus amongst stakeholders, a price has been determined and there is an incentive to do something. This is evident in biotechnology and health, and in communications. It is lacking in environmental concerns. Technology is an enabler; it facilitates advancement towards the goals.

Design

The ETR was widely accepted to be a policy with desirable effects, but design issues and political acceptability limited its implementation (Dresner, Dunne, Clinch *et al.*, 2006; Beuermann and Santarius, 2006). The polluter-pays principle underlies the vast majority of environmental taxes but becomes unsatisfactory in its implementation due to the design (or lack of design) of the taxes. Technology has failed to spur innovative environmental solutions because of inadequacies in its design and diffusion. Market failures, due to information deficiencies and inappropriate pricing, tend to stifle rather than stimulate the adoption of sustainable development technologies. And government regulation affects the innovation climate.

Environmental taxes to date have been individualistic, diffused and singularly focused. They are only combined as a category of taxation in revenue statistics. It is therefore not surprising that there are competing objectives and priorities and 'compromise' and 'trade-off' become synonymous with 'using economic principles to attain environmental goals'. It is therefore not surprising that corporate participation and acceptance levels are low. It is therefore not surprising that there is always a loser in environmental taxation.

Design features critical to participation and acceptance include providing adequate incentives, flexibility in delivery and ease of ongoing monitoring and compliance. But the necessary social, cultural and economic conditions are also required to foster an innovative climate and to provide the framework necessary to reward those corporate entities trying to be sustainable.

A corporate entity is a social construct, pursuing economic goals and accommodating environmental goals without conflict. If this balance is embedded in an environmental tax then there will only be winners. Therefore what drives corporate decision-making with respect to environmental concerns must be reflected in the design of the tax.

E. CONCLUSION

Climate change resilience demands measures that proactively increase productivity whilst reducing climate impacts. That is the challenge facing both business and government. However, addressing the environmental concerns of business has, to date, been governed by expediency, influence, pandering and confusion. The evidence portrays consistent economic failures in respect of environmental outcomes. Arguably, it exists that, singularly or collectively, no environmental taxation implementation has managed to deliver the professed outcome. Yes, they may have managed to add to government-consolidated revenue and some have managed to garner public support. However, the real environmental outcomes have been at best negligible and generally nonexistent or simply lost.

This pervasively negative paradigm has perpetuated the perspective that there is always a loser in environmental taxation. That is, any attempt to imbed or imbue sustainable practice will cost and not deliver the desired benefit. This is especially the case with business.

The norm that there must always be a loser should be challenged. The structure and approach to environmental taxation must enter a new paradigm – one where there is a direct economic incentive for business to adopt and develop sustainable business practices, government revenues are not adversely affected and there are substantive environmental outcomes. In other words, a win-win-win scenario.

NOTE

1. Also known as the 'millennium bug', a software problem that resulted from the practice of abbreviating a four-digit year to two digits.

REFERENCES

Ahdieh, R.B. (2009) 'The New Regulation: From Command to Coordination in the Modern Administrative State', www.works.bepress.com/robert_ahdieh/1/, accessed 31 August 2010.
Avi-Yonah, R. and D. Uhlmann (2009), 'Combating Global Climate Change: Why a Carbon Tax is a Better Response to Global Warming than Cap and Trade', *Stanford Environmental Law Journal*, **28**, 3.
Beuermann, C. and T. Santarius (2006), 'Ecological Tax Reform in Germany: Handling Two Hot Potatoes at the Same Time', *Energy Policy*, **34** (8), 917–29.
Bovenberg, A.L. and F. van der Ploeg (1998), 'Consequences of Environmental Tax Reform for Unemployment and Welfare', *Environment and Resource Economics*, **12** (2), 137–50.

Burritt, R.L., H. Lewis and K. James (2005), 'Analysing the Effectiveness of an Environmental Voluntary Agreement: The Case of the Australian National Packaging Covenant' in Edoardo Croci (ed.), *The Handbook of Environmental Voluntary Agreements: Design, Implementation and Evaluation Issues*, Dordrecht, The Netherlands: Springer.

Byers, D. (2010), 'Carbon Tax is Obviously Better Option', *Australian Financial Review*, 27 February 2010.

Couder, J. and A. Verbruggen, (2005), 'Towards an Integrated Performance Indicator for (Energy) Benchmarking Covenants with Industry' in Edoardo Croci (ed.), *The Handbook of Environmental Voluntary Agreements: Design, Implementation and Evaluation Issues*, Dordrecht, The Netherlands: Springer.

Deroubaix, J. and F. Lévèque (2002), 'The Design of Ecological Tax Reform: The French Ecotax' (PETRAS Final Report, Center of Industrial Economics, Ecole des Mines de Paris).

Dresner, S., L. Dunne, P. Clinch *et al.* (2006), 'Social and political responses to ecological tax reform in Europe: an introduction to the special issue', *Energy Policy*, **34** (8), 895–904.

Ellerman, D. (2000), 'Tradeable Permits for Greenhouse Gas Emissions: A primer with particular reference to Europe' (MIT Joint Program on the Science and Policy of Global Change, Report No 69).

Eurostat (2008), *Taxation trends in the European Union: Data for the EU Member States and Norway*, Luxembourg: European Communities.

Georg, S. (1994), 'Regulating the Environment: Changing from Constraint to Gentle Coercion', *Business Strategy and the Environment*, **3** (2), 11–20.

Grimeaud, D. (2000), 'The Integration of Environmental Concerns into EC Policies: A Genuine Policy Development?', *European Environmental Law Review*, **9** (7), 207–18.

Hepburn, C. (2010) 'Carbon Pricing: Is a Tax Better than Emissions Trading', Transcript of address given to the Grattan Institute, www.grattan.edu.au/assets/linked_docs/100325_Hepburn_Transcript.pdf, accessed 28 September 2010.

HM Revenue and Customs, 'Rates and Allowances Archive', www.hmrc.gov.uk/rates/archive.htm, accessed 22 September 2010.

Laskowska, A. and F. Scrimgeour (2002), 'Environmental Taxation: The European Experience' (Paper presented at the New Zealand Agricultural and Resource Economic Society Annual Conference).

Laves, L. and E. Malès (1989), 'At Risk: The Framework for Regulating Toxic Substances', *Environmental Science and Technology Journal*, **23** (4), 386–91.

Lawn, P. (2006), 'Ecological Tax Reform and the Double Dividend of Ecological Sustainability and Low Unemployment: An Empirical Assessment', *International Journal of Environment, Workplace and Employment*, **2** (4), 332–58.

Mann, I. (2009), *A Comparative Study of the Polluter Pays Principle and its International Normative Effect on Pollutive Processes*, British Virgin Islands: Forbes Hare.

Murphy C.J. (2002), *The Profitable Correlation Between Environmental and Financial Performance: A Review of the Research*, Seattle: Green Light Advisors.

National Research Council (2001), *Energy Research at DOE: Was it Worth it?*, Washington, D.C.: National Academy Press.

OECD (2002), *Technology Policy and the Environment*, Paris: OECD.

OECD (2003), *Voluntary Approaches for Environmental Policy: Effectiveness, Efficiency and Usage in Policy Mixes*, Paris: OECD.

OECD (2007), 'The Political Economy of Environmentally Related Taxes', *OECD Observer*, Paris: OECD.

OECD (2009), 'The Tax Treatment of Tradeable Permits for Greenhouse Gas Emissions', *Tax News Alerts*, Paris: OECD.

Pezzey, J. and A. Park (1998), 'Reflections on the Double Dividend Debate', *Environmental and Resource Economics*, **11** (3), 539–55.

Prasad, M. (2009), 'Taxation as a Regulatory Tool: Lessons from Environmental Taxes in Europe' in E Balleisen and D Moss (eds) *Toward a New Theory of Regulation*, Cambridge: Cambridge University Press.

Salzmann, O., A. Ionescu-Somers and U. Steger (2005), 'The Business Case for Corporate Sustainability: Literature Review and Research Options', *European Management Journal*, **23** (1), 27–36.

Schöb, R. (2003), 'The Double Dividend Hypothesis of Environmental Taxes: A Survey' (FEEM Working Paper No 60; CESifo Working Paper Series No 946).

United Nations (1992), 'Rio Declaration of Environment and Development', *Earth Summit Agenda 21: The United Nations Programme of Action from Rio*, Blue Ridge Summit, Pennsylvania: United Nations Publications.

Weaver, P., L. Jansen, G. van Grootveld *et al.* (2000), *Sustainable Technology Development*, Sheffield: Greenleaf.

Williams, D.F. (2008), 'Taxation and the Environment' (A discussion paper, KPMG's Tax Business School, UK).

6. Behavioural strategies to support climate change resilience

Amanda Kennedy and Wanida Phromlah

INTRODUCTION

Environmental taxation reform is often intended to provide a strong incentive for adopting sustainable behaviours that can assist in achieving climate change resilience, such as Pigouvian taxes on energy use and other price mechanisms (Martin and Werren 2009, p. 1). Arguably, these are often the 'first best' instrument to support climate change resilience. Yet such 'first best' instruments are often abandoned, with governments demonstrating a limited appetite for these mechanisms in the light of high political transaction costs (Martin and Werren 2009).

A failure to implement the recommendations of the recent Henry Review of Australia's taxation system (which also incorporated prior proposals for a carbon emissions trading scheme) provides a clear illustration of how the political economy can frustrate a policy package. This has left Australia with no specific economic mechanisms to deal with climate change since implementation of the preferred carbon pricing scheme has also been delayed. In the absence of a credible market-based instrument focused upon climate change, the 'first best' instrument choice seems not to be immediately available.[1] In light of this, what can be done to support climate change resilience? And, even if the 'first best' instrument were adopted, what tactical support would make it most likely to succeed, particularly given that any such scheme is embedded in complex multi-party transacting systems?

We argue that non-market behavioural strategies are likely to be of great potential significance, either when 'first best' instruments are defeated, delayed or compromised; and/or even when 'first best' instruments are endorsed, to provide the support that is necessary for effective implementation. In this chapter, we examine the role that behavioural tools, like information and social stimuli, may play as an alternative or complementary option to taxation and regulation. We argue that economic policy that focuses on the central, modelled ideal instrument is somewhat myopic, as it masks what is needed to achieve the underlying goals of the instrument. We propose that the systemic use of complementary behavioural tools will often be important to efficiently shift patterns of resource consumption towards more sustainable levels, and may even operate as central mechanisms where taxation and regulatory approaches are either absent or inadequate. We focus on the potential for the use of total system strategies, within which market and regulatory arrangements may be a centrepiece but not the totality of the strategy. For the purpose of this discussion emphasis is given to the use of community-based social marketing. Community-based social marketing focuses upon removing barriers to behavioural change, and promoting positive change via a suite of interventions founded in social psychology.

The chapter begins with a recapitulation of the literature relevant to the behavioural effectiveness of environmental taxation instruments, before turning to an analysis of community-based social marketing as a complementary behavioural strategy to support economic instruments for climate change resilience.

THE BEHAVIOURAL EFFECTIVENESS OF ENVIRONMENTAL TAXATION INSTRUMENTS

This chapter does not seek to revisit the extensive literature on the advantages (and disadvantages) of pursuing environmental taxation as a 'first best' means to reduce environmental harm. Suffice to say that there are ample illustrations of environmental taxation instruments shifting patterns of resource consumption towards more sustainable levels, as well as an engagement of the debate between 'tax versus no tax' as a means to achieve behavioural outcomes.[2] Of greater concern here is the behavioural effectiveness of environmental taxation in situations where taxation instruments suffer from institutional impediments, or where they are otherwise delayed or not pursued.

i. The Political Economy

A limited political 'appetite' (Martin and Werren 2009, p. 1) for environmental taxation reform means in some situations proposed environmental taxation instruments are left abandoned, often due to the potential political transaction costs of implementing such mechanisms. A recent example from Australia highlights this. The *Australia's Future Tax System* report, released in 2010, detailed many recommendations for improving Australia's tax system. One priority of the review was to improve the environmental performance of Australia's tax system. The report acknowledged that environmental taxes are a key option open to government to address 'spillover' costs from market failure upon the environment (Commonwealth of Australia 2010, p. 343), endorsing the then-proposed national Carbon Pollution Reduction Scheme (CPRS) and introducing specific environmental taxation reforms to support the proposed CPRS.

By the time the government released its response to the Henry Review in May 2010, the CPRS had been shelved until 2013 (after the expiry of the Kyoto Protocol) (Kelly 2010), and as a result, none of the reform recommendations were adopted (indeed, only a handful of the 138 recommendations from the Henry Review were 'cherry picked' for implementation) (Maiden 2010). The surrounding political climate in Australia at the time of the Henry Review must also be acknowledged when considering the ultimate rejection of Henry's proposed 'first best' instruments. When the terms of reference were initially discussed for the Henry Review, it was indicated that one of its overarching objectives was to realign taxation across a range of areas with sustainability (particularly the proposed emissions trading scheme).[3] The eventual release of the Henry Review saw the inclusion of a range of recommendations intended to work alongside the proposed CPRS to align taxation with sustainability. Ultimately, the government decided to implement a 40 per cent resources sector super profits tax which was only loosely aligned to the approach proposed by the Henry Review, and did not introduce any other environmental taxation reforms. Coupled with the limited opportunity for consultation over the Henry recommendations, this reform attracted a great deal of public criticism, effectively derailing any 'first best' policy responses before they were introduced (Maiden 2010).

Shortly after the response to the Henry Review was released, there was a leadership coup within the governing political party (Rodgers 2010), followed by a federal election that resulted in the first hung parliament in Australia in 70 years (Liddy 2010). In the lead-up to the election, a range of policies were proposed by both parties which fell short of the combination of the CPRS and environmental taxation reform advocated by the Henry Review. Since the August 2010 election, the newly installed government has stated that it will move towards a full review of the Australian tax system, and has indicated that a carbon tax is back on the agenda for discussion (Harper 2010; Salna 2010). The original resources sector super profits tax proposal is also being re-negotiated as a minerals resource rent tax, which is being met with some scepticism in light of the composition of the government consultation group, which includes several large mining companies (Hewett 2010).

This brief summary of the political dynamics of the Australian Henry Review demonstrates the influence of the political economy on the uptake of 'first best' instruments, consistent with public choice theory (Park, Conca, and Finger 2008; Schroeder 2009). On its most basic construction, public choice theory holds that government operates in a system where participants 'ignore welfare-improving actions in favor of ones that advance their own narrow self-interests, and where participants representing economically powerful special interests predominate' (Schroeder 2009, p. 1). As a result, government decisions benefit industry at the expense of any wider public concerns of environmental standards (Schroeder 2009; Stigler 1971). However, Schroeder cautions against the use of simple models of public choice theory to explain government decisions, as self-interest may not necessarily drive all participants within the system, including government (Schroeder 2009). Human behaviour is not necessarily rational, and may depend upon the market setting in which a decision is made. As Schroeder argues, the:

> ... shift from thick rationality to thin rationality assumptions about the behavior of voters, legislators and bureaucrats better aligns public choice's behavioral assumptions with those that market theory applies to market actors. Acknowledging the possibility that political actors can be motivated by the prospect of solidary or purposive benefits and by principled commitments in addition to material self interest also complicates the analysis. (Schroeder 2009, p. 31)

Whilst self-interest is a 'powerful motivator' for government behaviour, the influence of all competing benefits ought be assessed, including the role of ideological commitments and the possibility of government making a decision in the broader public interest (Schroeder 2009). Consistent with this, several scholars have in more recent times acknowledged the whole of the surrounding political economy in seeking to explain the state of environmental governance (Eliadis, Hill, and Howlett 2005; Park, Conca, and Finger 2008; Stavins 2004), examining a range of influences upon the setting of environmental standards including the weighting of costs and benefits by agencies, the role of equity concerns, and the sway of powerful interest groups (Stavins 2004). The recent situation in Australia illustrates a scenario where multiple complex influences have shaped the surrounding political economy, ultimately resulting in the rejection of 'first best' instruments. Minimally, this literature and the Australian example point to the practical necessity of designing strategies that are able to withstand the vagaries of politics.

ii. Instrument Choice

Setting aside the influence of political agency leading to the non-adoption or distortion of preferred instruments, even where 'first best' instruments are feasible, reliance on one instrument and one point of intervention is not likely to be optimal in dealing with complex multi-party transacting systems. It is therefore useful to consider complements and supports for the implementation of economic instruments. Whilst the Henry Review notes that there 'is no single instrument suitable for all environmental issues', the report notes that in general, single policy instruments should be used for each separate environmental objective, and that multiple instruments are inefficient and costly and should only be considered where a single instrument is incapable of achieving the desired goal. We question this perspective, for it seems that this ignores the systems nature of resource use, and also the proven value of tactical interventions to support introduction of change. As Eliadis, Hill, and Howlett (2005, p. 7) note:

> Knowledge about how instruments perform relative to each other, especially in different governance contexts, is generally still limited. Public servants often complain that they are ill-equipped to assess the relative merits of instrument choices and that far more knowledge is needed, especially with respect to the effectiveness of 'innovative' instruments and instrument mixes such as voluntary codes, partnerships and co-regulatory instruments. Basic research questions remain unanswered, including which instruments are most likely to work, either singly or in combination, and how instruments interact both with public and private actors and institutions and with past policy choices.

It would seem evident that a form of instrumental myopia is preventing strategists from fully considering economic, regulatory and voluntary tools to be implemented simultaneously which might better assist in supporting climate change resilience. It has been argued that the single instrument focus 'is especially lamentable because it has afflicted the study and practice not only of regulation, but also of the other tools' (Eliadis, Hill, and Howlett 2005, p. 7). A focus on the implementation of a single instrument to the exclusion of other complementary instruments is problematic if the chosen instrument is not adopted, compromised, or where the chosen instrument is adopted but fails to achieved the desired behaviour change.[4] Faure argues that 'there is not just one optimal instrument of environmental policy ... the key issue is how one can find an optimal combination of various instruments to reach environmental goals at the lowest cost' (Faure 2009, p. 17). Faure (2009) goes on to suggest that the empirical data on market instruments shows that it is in fact rare for a market instrument alone to lead to improved environmental outcomes. A strategy is required which uses other instruments to 'fill the gap' until such a time when suitable economic instruments are introduced, and also as a complement to economic instruments on their implementation.

Whilst a variety of studies of instrument choice have focused on the merits of single instruments, the next generation of studies on instrument choice ought deal with the fact that 'governments achieve public purposes by "steering" diverse, complex networks of public and private actors, institutions, ideas, and policy instruments' (Eliadis, Hill, and Howlett 2005, pp. 4–5). As Lin notes, efforts to change behaviour with respect to the environment:

> ... should not rely exclusively on any single approach such as voluntary behavioral changes or strict mandates. The breadth and severity of the problem demand the deployment of a range of policy tools, including traditional forms of regulation, economic incentives, information-based regulation, education, and voluntary efforts. Numerous factors affect behavior, and the effectiveness of any particular tool for bringing about behavioral change will depend on context. (Lin 2009, p. 1148)

How then might we blend concern with the choice of instrument within a broader strategy for total behaviour change? Is it possible to move beyond pitting taxation against other instruments to craft a total system strategy using multiple instruments applied at different points to manage transaction behaviour (Martin and Verbeek 2000)?

Behavioural economics has attempted to integrate psychological concepts with economic perspectives to better understand how actors behave and respond to policy interventions in markets (McAuley 2010). McAuley (2010) provides a detailed summary of the development of this field, and advocates increased use of findings from empirical studies in behavioural economics to inform policy. Scholars of behavioural economics have recognised that government confidence in passive market mechanisms (such as pricing) do not take into account 'the behavioral biases which get in the way of wise decision-making' (McAuley 2009, p. 16). Indeed, as Australian economist Ross Gittins remarked recently, economists 'tell themselves government regulation fails because of the corruptibility of politicians and public servants. It never occurs to them the root cause may be their own shaky grasp on human motivation' (Gittins 2010).

There has also been an increased focus in the regulatory literature upon behavioural concepts, and in particular the use of complementary mechanisms within regulatory strategy to achieve behavioural change (Vandenbergh 2005; Vandenbergh and Steinemann 2007; Dernbach 2007).

Voluntary and quasi-voluntary mechanisms are gaining traction as complementary instruments which might assist regulation as a part of a total systems strategy. Some of the regulatory literature has recognised that more effective behavioural outcomes may be gained where regulation is combined with voluntary tools designed to stimulate positive motivations for compliance (Gunningham and Grabosky 1998; Eliadis, Hill, and Howlett 2005). Concepts of 'smart' (Gunningham and Grabosky 1998) and 'responsive' (Ayres and Braithwaite 1992) regulation are being advocated, which involve a carefully structured mix of regulatory and non-regulatory mechanisms. In particular, the corporate arena provides abundant examples of credible 'self-regulatory' approaches which promote moving 'beyond compliance' through such initiatives as triple bottom line reporting, eco-auditing and corporate social responsibility programs (Feldman 2009; Graafland and van de Ven 2006; Campbell 2007; Esty and Winston 2006).

With regulators and behavioural economists increasingly more willing to look to complementary instruments to strengthen the behavioural effectiveness of certain regulatory and economic tools, might not economic strategists consider the role of supplementary tools as a part of a total system to achieve desired behavioural outcomes? Analysis of the effectiveness of various standalone economic and regulatory approaches is abundant in the literature. What has not yet received sufficient attention is the interaction between these instruments and behavioural tools founded in psychological studies to improve environmental behavioural outcomes (McAuley 2010; McAuley 2009). In the balance of this chapter we will demonstrate the potential with community-based social marketing as an example. This approach promotes a structured program that focuses first on removing barriers to behaviour change, and using specific behaviour change tools at a direct community level to elicit change (McKenzie-Mohr and Smith 1999). Whilst such approaches are subject to limitations, recent research has explored the potential for these tools to complement regulatory programs (Becker *et al.* 2010; Kennedy 2010). The remainder of this chapter explores the possibility of combining community-based social marketing programs with economic instruments to help change environmental behaviour, offering fresh perspectives for future environmental taxation research.

COMMUNITY-BASED SOCIAL MARKETING AS A COMPLEMENTARY MECHANISM FOR ECONOMIC INSTRUMENTS

Community-based social marketing has developed as a distinct field from social marketing, the latter of which is focused upon creating public awareness of an issue. Social marketing tends to employ information advertising alone to create awareness of issues, whereas community-based social marketing focuses on behaviour change, utilizing techniques drawn from social psychology (McKenzie-Mohr and Smith 1999). It has a strong focus on first identifying barriers to behaviour change, then applying specific tools and techniques to encourage a desired behavioural change (McKenzie-Mohr and Smith 1999).[5] Whilst information advertising is a part of the community-based social marketing toolkit, other tools are also used to help transcend 'the gap between knowledge [and] action' (Kollmuss and Agyman 2002, p240) to achieve sustainability outcomes.

The underlying philosophy of community-based social marketing is that attempts to change behaviour are more effective when delivered at the local scale using direct means of contact and interaction (McKenzie-Mohr and Smith 1999). Gaining a deep understanding of the barriers to behaviour change is advocated, before using behavioural tools to overcome these barriers (McKenzie-Mohr and Smith 1999). It is difficult to change behaviour where barriers exist to sustainable alternatives (Dernbach 2007; Pollard 2005), and in some cases, it may be necessary to increase the benefits of adopting desired behaviours, or remove obstacles to their adoption (Dernbach 2007). Once the barriers to engaging in behaviour change are understood, tools may be implemented to overcome these barriers. The tools presently include using effective communication strategies, seeking commitments to change behaviour, using prompts to remind people about behaviour change (and the benefits of change), development and reinforcement of community norms, and use of appropriate incentives (McKenzie-Mohr and Smith 1999). It is also advocated in community-based social marketing programs to conduct a pilot program to test the mechanisms before they are implemented on a wider scale, as well as to conduct regular monitoring and evaluation of programs to determine whether the strategy needs refinement (McKenzie-Mohr and Smith 1999).

The tools prescribed in the community-based social marketing toolkit are supported by a growing body of empirical research particularly about improving individual environmental behaviour. Whilst every behavioural system is complex, and the barriers to adoption of sustainable behaviours will depend upon a range of contextual factors (Martin and Verbeek 2006), the examples discussed below offer guidance as to the potential tools available.

Building upon traditional marketing theory, community-based social marketing approaches include the use of effective *communication* techniques (McKenzie-Mohr and Smith 1999). To be effective, communication regarding the adoption of a desired behaviour change must be personal, targeted, simple, and delivered by a trusted source (McKenzie-Mohr and Smith 1999).[6] In an analysis of the application of social psychology tools to energy consumption, Aronson (1990) found that vivid communication strategies that captured the attention of the participant were more successful at bringing about behaviour change in energy use audits. This included comparing the size of cracks around doors to relative objects (such as a basketball), detailing the dollar costs of not implementing improvements, and involving the participant in the audit inspection. This program had a good success rate, with an estimated 60 per cent of participants implementing energy efficient home improvements as a result of their participation in the audit program.

Seeking *commitment* is another tool in the community-based social marketing toolkit, whereby individuals are approached to pledge their commitment to a particular action (McKenzie-Mohr and Smith 1999). As McKenzie-Mohr and Smith (1999) note, the process of building commitment is often underestimated as an effective psychological tool. Quite remarkable results have been reported from various studies, owing to the fact that individuals like to be seen as behaving consistently (McKenzie-Mohr and Smith 1999). If an individual has 'given their word' that they will do something, or that they will support a particular cause, they are unlikely to behave inconsistently with this (Cialdini 1993). Where such commitments are given publicly, this serves to strengthen an individual's resolve. Pallak, Cook and Sullivan's (1980) investigation of the use of commitments to reduce energy consumption illustrates just how effective this tool can be. In their research, use of natural gas and electricity in Iowa City was assessed during home visits, and householders were provided with information about conservation and energy reduction strategies. A public commitment to reduce energy was sought from one group of participants, who were told that their names would be published with the results of the program. From this group, a reduction of approximately 10–20 per cent in energy consumption was observed, compared with no significant decrease in the control group (Pallak, Cook and Sullivan 1980).

Prompts, whilst not usually sufficient to generate attitudinal change in their own right, may nonetheless assist in reminding individuals to engage

in sustainable behaviours that they have otherwise agreed to (i.e. remember-
ing recyclable shopping bags) (McKenzie-Mohr and Smith 1999).
McKenzie-Mohr and Smith (1999) note that prompts need to be self-
explanatory and noticeable – they need to be located in close proximity to
the target behaviour. Austin *et al.* (1993) demonstrate how effective
prompts can assist in improving uptake of desired behaviours in a campus
recycling program. In a seemingly simple experiment, it was found that in
recycling activity increased by 54 per cent in a department where information about recycling (including information about separation of trash and
recyclables) was placed in immediate proximity to recycling receptacles.
This was compared with much smaller increases in a second department,
where recycling information was placed several meters away from the
receptacle (which increased after the information prompts were moved
closer to the recycling receptacle).

Social norms are another key feature of the community-based social
marketing toolkit, the development of which may be used to guide behaviour backed by social sanctions (McKenzie-Mohr and Smith 1999;
Vandenbergh and Steinemann 2007; McAuley 2010). Individuals measure
their behaviour against their perceptions of 'the norm', thus social norm
development aims to communicate that undesirable behaviours are not
occurring frequently, thereby reducing the possibility that such behaviours
become accepted practice (Schultz *et al.* 2007). Schultz *et al.* (2007) assessed
the role of social norms in reducing energy consumption in California. In
their study, households were provided with normative information about
energy consumption, and then monitored for a period of four weeks. At the
commencement of the monitoring period, households were provided with
door hangers that displayed information about the household energy use
over the previous week, along with normative information about the
average energy consumption in the neighbourhood. In the group of participants who used less energy than the neighbourhood average, half were
provided with information about reducing energy consumption, and the
other half were also provided with this information as well as a 'smiling
face' icon on their hanger (to indicate approval of their rate of energy
consumption). In the group of participants who used more than average
energy, again half were provided with energy consumption reduction
information only, and the other half were provided with the information as
well as a 'frowning face' icon on their door hanger to indicate disapproval
with their rate of consumption. It was subsequently observed in the group
who received the frowning face icon that electricity consumption was
reduced by 6 per cent, compared with 4.6 per cent in the other half of the
group who used more energy than the average. In the group who used less
electricity than average, the half who received the smiling face icon

increased their energy consumption by 1 per cent, compared with 10 per cent for the half of the group that did not receive the normative support via the smiling face approval. It was concluded that this minor act of providing normative information encouraged those who consumed more energy than average to reduce consumption, and motivated those who were already consuming less energy than the average to maintain their level of use (Schultz *et al.* 2007).

The final tool articulated in the community-based social marketing toolkit is the use of *incentives*. Incentives are already known within economics for their ability to generate behaviour change, and similar observations are evident from social psychology (McKenzie-Mohr and Smith 1999; McAuley 2010). It has been noted that incentives are most effective when they are positive (rather than a disincentive, such as user fees for garbage collection which might prompt illegal dumping) (McKenzie-Mohr and Smith 1999). Incentives must also be paired with the behaviour change, and visible. Wang and Katzev (1990) illustrate how the community-based social marketing tools of incentives and commitment may be used to foster sustainable behaviours. In the recycling program they observed, a group who made a public commitment to recycle increased their recycling practices by 48 per cent, and a second group who were offered coupons from local businesses as an incentive to recycle increased their recycling by 54 per cent.

DISCUSSION

Much economic policy analysis is focused on finding the optimal single instrument to address a defined issue. Often that instrument will be a taxation or pricing instrument. However, in the real world of implementation, the ideal and what happens are often far apart. Rejection, distortion and delay in the implementation of 'first best' instruments seems to be the norm rather than the exception. Further, even when the 'first best' instrument is adopted as a core architecture to change collective behaviour, there are many micro causes of delay and compromise to the occurrence of the transactions that the economic interest presupposes will occur.

Whilst the examples listed in the previous section are few, they illustrate the core components of the community-based social marketing toolkit. They illustrate some of the potential complementary methods available to assist in generating behaviour change, particularly where combined with economic or regulatory mechanisms. Perhaps one of the leading examples of a comprehensive community-based social marketing program comes from Portland, Oregon in the USA where a whole of community strategy was implemented to assist in meeting legislated air pollution standards.[7] Metropolitan Portland had failed to meet the standards of air quality under the Clean Air Act, which prompted the Department of Environmental Quality (DEQ) to initiate an 'Air Quality Public Education and Incentive Program'. Using a range of tools advocated by the community-based social marketing movement, the DEQ (in conjunction with community and enterprise partnerships) targeted activities which would impact upon the emission of volatile organic compounds (VOCs), including vehicle use, lawnmower use, and certain paint products. The community-based social marketing tools used included prompts (through in-store promotion of low-VOC emission products by shelf signage, staff buttons and regular information sessions); incentives (discounts for low VOC products, buy-back schemes for petrol lawn mowers and rebates for electric lawn mowers, and retail vouchers for reducing non-work vehicle use); commitment (seeking pledges from residents to find alternate methods of transport for non-work trips, or adopting car-pooling); development of norms (instilling community wide 'Clean Air Action Days' when smog levels were high, where residents and businesses voluntarily reduced VOC emitting activities); and effective communication strategies to promote all aspects of the program. This community-based social marketing program resulted in Oregon being recategorised as having attained the Clean Air Act standards, and illustrates the potential for complementary tools to enhance other approaches, such as regulation and market tools, to behaviour change.

As Gittins (2010) remarked recently, economists 'specialise in studying the behaviour of markets. But economic sociologists know a lot about how markets work – that they're social phenomena, influenced by the informal constraints of social norms as well as the formal constraints of public policy'. What may be needed to tap into these social phenomena are alternative or complementary micro-level behaviour change strategies. We have illustrated some of these possibilities, with an emphasis on community-based social marketing.

CONCLUSION

As a part of a broad strategy to change environmental behaviour to achieve climate change resilience, the adoption of appropriate community-based social marketing tools may strengthen market instruments that are designed to change behaviour. Where economic approaches are inadequate or absent, the behaviour change tools offered by community-based social marketing might offer an alternative means of changing behavioural practices. In some cases, they may even be better suited than economic tools, particularly where such instruments do not attack the right behavioural drivers. The available literature and case study evidence suggests real potential for both regulatory and economic instruments to be more effective when they are considered strategically alongside other complementary tools as a part of a broader initiative to elicit behaviour change. Stern (2005) argues that the 'record of single-strategy approaches to changing consumer behavior is, in short, mixed at best', and community-based social marketing offers a substantial range of tools which can be combined with regulatory and economic mechanisms to achieve desired behaviour change.

What are the implications of these considerations? We would argue that the gap between theoretical potential and realised outcome from economic instruments may be partly bridged by far greater attention to the micro-level implementation strategies for change than has hitherto been the norm. Economics has become increasingly more concerned with behaviour issues, and this chapter reinforces the need for this concern and suggests substantial practical value from doing so. As Eliadis, Hill and Howlett (2005) note, instruments are context sensitive and are rarely designed or indeed implemented in isolation. It is time for the 'instrument choice' debate to more actively consider elegant instrument mixes, taking into account the surrounding contextual institutional and cultural factors of the target market.

This chapter also suggests that the potential value of a greater fusion between different fields of scholarly endeavour. Scholars in economics, regulation and behaviour are all focusing in their different fields on the same basic question: 'How can we achieve pro-sustainability behaviour change at least cost?' However, cross-fertilization and shared learning seems to be limited. Perhaps overcoming this problem is the next frontier for a variety of disciplines.

NOTES

[1.] Though note that the recently elected government has indicated a renewed commitment to carbon pricing in some as yet undetermined manner.

[2.] See generally the previous volumes of *Critical Issues in Environmental Taxation*, including Milne (2003); Cavaliere (2006); Deketelaere (2006) and (2007); Chalifour, Milne, Ashiabor, Deketelaere, and Kreiser (2008); Cottrell (2009); and Lye, Milne, Ashiabor, *et al.* (2009). See also Harrington, Morgenstern, and Sterner (2004); Faure (2009); and Morley (2010), where a comprehensive empirical analysis of environmental taxes across a panel of EU members and Norway illustrated a 'significant negative relationship between taxes and pollution'.

[3.] See, for example, http://taxreview.treasury.gov.au/content/Content.aspx?doc=html/reference.htm.

4. For example, Morley (2010) has identified several instances where environmental taxes imposed in the European Union have not resulted in shifting patterns of behaviour with respect to energy consumption.
5. For a more detailed history of the development of community-based social marketing from social marketing, see Kennedy (2010).
6. See also McAuley (2009), who notes that concrete terms are better than abstract terms in conveying messages to consumers: 'A message such as "check your front door for drafts" is more likely to promote action than one such as "make sure your house is well sealed".', p. 10.
7. See Tools of Change, 'Oregon's Air Quality and Public Education Program', www.toolsofchange.com/en/case-studies/detail/101/, accessed 1 October 2010.

REFERENCES

Aronson, E. (1990), 'Applying social psychology to desegregation and energy conservation', *Personality and Social Psychology Bulletin* (Special Issue: Illustrating the Value of Basic Research), **16** (1), 118–32.

Austin, J., D.B. Hatfield, A.C. Grindle, *et al.* (1993), 'Increasing recycling in office environments: The effects of specific, informative cues', *Journal of Applied Behavior Analysis*, **26** (2), 247–53.

Ayres, I. and J. Braithwaite (1992), *Responsive Regulation: Transcending the Deregulation Debate*, New York: Oxford University Press.

Becker, J.C., A.E. Luloff, J.C. Finley, *et al.* (2010), 'Towards a contemporary behavioural science basis for effective regulation', paper presented at the *Eighth Annual IUCN Academy of Environmental Law Colloquium*, 13–17 September 2010, Ghent, Belgium.

Campbell, J.L. (2007), 'Why would corporations behave in socially responsible ways? An institutional theory of corporate social responsibility', *Academy of Management Review*, **32**, 946–67.

Cavaliere, A. (ed.) (2006), *Critical Issues in Environmental Taxation Volume III: International and Comparative Perspectives*, Oxford: Oxford University Press.

Chalifour, N., J.E. Milne, H. Ashiabor, *et al.* (eds.) (2008), *Critical Issues in Environmental Taxation Volume V: International and Comparative Perspectives*, Oxford: Oxford University Press.

Cialdini, R.B. (1993), *Influence: Science and Practice*, New York: HarperCollins.

Commonwealth of Australia (2010), *Australia's Future Tax System: Report to the Treasurer*, Part Two: Detailed Analysis, Volume 2 of 2.

Cottrell, J. (ed.) (2009), *Critical Issues in Environmental Taxation Volume VI: International and Comparative Perspectives*, Oxford: Oxford University Press.

Dernbach, J.C. (2007), 'Harnessing individual behavior to address climate change: options for Congress', *Virginia Environmental Law Journal*, **26**, 107–57.

Deketelaere, K. (ed.) (2006), *Critical Issues in Environmental Taxation Volume II: International and Comparative Perspectives*, Oxford: Oxford University Press.

Deketelaere, K. (ed.) (2007), *Critical Issues in Environmental Taxation Volume VI: International and Comparative Perspectives*, Oxford: Oxford University Press.

Eliadis, P., M. Hill, and M. Howlett (eds.) (2005), *Designing Government: From Instruments to Governance*, Montreal: McGill-Queen's University Press.

Esty, D. and A. Winston (2006), *Green to Gold: How Smart Companies Use Environmental Strategy to Innovate, Create Value, and Build Competitive Advantage*, New Haven, Connecticut: Yale University Press.

Faure, M. (2009), 'Instruments for environmental governance: what works?', paper presented at the *Seventh Annual IUCN Academy of Environmental Law Colloquium*, 2009, Wuhan, China.

Feldman, I.R. (2009), 'Business and Industry: Transitioning to Sustainability', in J. Dernbach (ed.), *Agenda for a Sustainable America*, Washington, D.C: Environmental Law Institute Press, 71–91.

Gittins, R. (2010), 'Psychologists needed to help implement policies', *The Sydney Morning Herald*, October 4, 2010.

Graafland, J., and B. van de Ven, B. (2006), 'Strategic and moral motivation for corporate social responsibility', *Journal of Corporate Citizenship*, **22**, 111–23.

Gunningham, N. and P. Grabosky (1998), *Smart Regulation: Designing Environmental Policy*, New York: Oxford University Press.

Harper, A. (2010), 'Fraser not harmonising with Swan', *The Sydney Morning Herald*, http://news.smh.com.au/breaking-news-national/fraser-not-harmonising-with-swan-20101004–1645w.html, 4 October 2010, accessed 5 October 2010.

Harrington, W., R.D. Morgenstern, and T. Sterner (eds.) (2004), *Choosing environmental policy. Comparing instruments and outcomes in the United States and Europe*, Washington: Resources for the Future.

Hewett, J. (2010), 'Explosive government resource tax paper', *The Australian*, http://www.theaustralian.com.au/business/opinion/explosive-government-resource-tax-paper/story-e6frg9if-1225933013247, accessed 2 October 2010.

Kelly, J. (2010), 'Kevin Rudd delays emissions trading scheme until Kyoto expires in 2012', *The Australian*, www.theaustralian.com.au/national-affairs/climate/kevin-rudd-delays-emissions-trading-scheme-until-kyoto-expires-in-2012/story-e6frg6xf-1225858894753, 27 April 2010, accessed 2 October 2010.

Kennedy, A.L. (2010), 'Using community-based social marketing techniques to enhance environmental regulation, *Sustainability*, **2**, 1138.

Kollmuss, A. and J. Agyeman (2002), 'Mind the Gap: why do people act environmentally and what are the barriers to pro-environmental behavior?' *Environmental Education Research*, **8** (3), 239–60.

Liddy, M. (2010), 'Australia's hung parliament explained', *ABC News Online*, http://www.abc.net.au/news/stories/2010/08/23/2990782.htm, 23 August 2010, accessed 2 October 2010.

Lin, A.C. (2009), 'Evangelizing climate change', *New York University Environmental Law Journal*, **17**, 1135–93.

Lye, L.H., J.E. Milne, H. Ashiabor, *et al.* (eds.) (2009), *Critical Issues in Environmental Taxation: International and Comparative Perspectives Volume VII*, Oxford: Oxford University Press.

McAuley, I. (2010), 'When does behavioural economics really matter?', paper presented at the *Australian Economic Forum*, Sydney, 5–6 August 2010.

McAuley, I. (2009), 'Carbon and consumers', paper presented at the *Carbon and Consumers Conference*, Public Interest Advocacy Centre, Sydney, February 2009.

McKenzie-Mohr, D. and W. Smith (1999), *Fostering Sustainable Behavior: An Introduction to Community-Based Social Marketing*, Gabriola Island: New Society Publishers.

Maiden, M. (2010), 'With election looming, government opts to cherry pick from Review', *The Age*, www.theage.com.au/business/with-election-looming-government-opts-to-cherry-pick-from-review-20100502-u13c.html, 2 May 2010, accessed 2 October 2010.

Martin, P. (2008), 'The Changing Role of Law in the Pursuit of Sustainability', in M. Jeffery, J. Firestone, and K. Bubna-Litic (eds.), *Biodiversity, Conservation, Law and Livelihoods: Bridging the North-South Divide*, New York: Cambridge University Press, 49–65.

Martin, P. and M. Verbeek (2006), *Sustainability Strategy*, Sydney: The Federation Press.

Martin, P. and M. Verbeek (2000), *Cartography for Environmental Law: Finding New Paths to Effective Resource Regulation*, Sutherland, Australia: The Profit Foundation.

Martin, P. and K. Werren (2009), 'The Use of Taxation Incentives to Create New Eco-Service Markets', in L.H. Lye, , J.E. Milne, H. Ashiabor, *et al.* (eds.), *Critical Issues in Environmental Taxation: International and Comparative Perspectives Volume VII*, Oxford: Oxford University Press.

Milne, J.E. (ed.) (2003), *Critical Issues in Environmental Taxation Volume I: International and Comparative Perspectives*, Oxford: Oxford University Press.

Morley, B. (2010) 'Empirical Evidence on the Effectiveness of Environmental Taxes', *Working Paper*, Department of Economics, University of Bath, http://opus.bath.ac.uk/18105/1/0210.pdf , accessed 3 October 2010.

OECD (2005), *Survey of Trends in Taxpayer Service Delivery Using New Technologies*, Centre for Tax Policy and Administration, Forum on Tax Administration, Paris: OECD.

Pallak, M.S., D.A. Cook, and J.J. Sullivan (1980), 'Commitment and Energy Conservation', in L. Bickman (ed.), *Applied Social Psychology Annual*, Beverley Hills: Sage, 235–53.

Park, J., K. Conca, and M. Finger (2008), *The Crisis of Global Environmental Governance: Towards a New Political Economy of Sustainability*, Oxford: Routledge.

Pollard, T. (2005), 'Driving change: public policies, individual choices, and environmental damage', *Environ. Law Rep.*, **35**, 10791−9.

Prime Ministerial Task Group on Emissions Trading (2007), *Report of the Task Group on Emissions Trading*, Canberra: Department of the Prime Minister and Cabinet.

Rodgers, E. (2010), 'Gillard ousts Rudd in bloodless coup', *ABC News Online*, www.abc.net.au/news/stories/2010/06/24/2935500.htm, 24 June 2010, accessed 2 October 2010.

Salna, K. (2010), 'Carbon tax no certainty: Crean', *The Sydney Morning Herald*, http://news.smh.com.au/breaking-news-national/carbon-tax-no-certainty-crean-20101003-162dt.html, 3 October 2010, accessed 3 October 2010.

Schroeder, C.H. (2009), 'Public Choice and Environmental Policy: A Review of the Literature', *Duke Law School Faculty Scholarship Series*, Paper 175, http://lsr.nellco.org/duke_fs/175, accessed 2 October 2010.

Schultz P.W., J.M. Nolan, R.B. Cialdini, *et al.* (2007), 'The constructive, destructive, and reconstructive power of social norms', *Psychological Science*, **18** (5), 429–34.

Stavins, R.N. (ed.) (2004), *The Political Economy Of Environmental Regulation*, UK: Edward Elgar Publishing.

Stern (2005), 'Understanding individuals' environmentally significant behaviour', *Environ. Law Rep.*, **35**, 10785–90.

Stigler, G.J. (1971), 'The theory of economic regulation', *Bell Journal of Economics*, **2** (1), 3–21.

Tools of Change, 'Oregon's Air Quality and Public Education Program', www.toolsofchange.com/en/case-studies/detail/101/, accessed 1 November 2010.

Vandenbergh, M.P. (2005), 'Order without social norms: how personal norm activation can protect the environment', *North Western University Law Review*, **99**, 1101.

Vandenbergh, M.P. and A.C. Steinemann (2007), 'The carbon neutral individual', *New York University Law Review*, **82**, 1673.

Wang, T.H. and R.D. Katzev (1990), 'Group commitment and resource conservation: two field experiments on promoting recycling', *Journal of Applied Social Psychology*, **20** (4), 265–75.

PART III

Environmental Taxation and Land
Management

7. Taxing land rents for urban livability and sustainability

H. William Batt

INTRODUCTION

In most cities of the world today ambience and livability are plagued with two problems: traffic congestion and sprawl development. Yet public bodies seem at a loss in solving them, even though at least from a technical point of view they are demonstrably solvable. Governments have at their command two means by which to address them – two arrows in their quiver, so to speak: constitutionally known as police powers and tax powers.[1] More commonly referred to as command-and-control approaches and fiscal approaches, they are the only legitimate tools that the public has at its disposal.

All this must be borne in mind when designers of government policy consider the efficacy of public programs, particularly with reference to their scope, domain, and weight. Scope involves all those matters or interests in which government concerns itself; the domain is the area or number of people over which it has exercise; and the weight, or intensity, is the degree to which a people or an area feels itself imposed upon, heavily or only lightly. If a government in some way over-extends itself, or imposes itself too much upon people, it will prove to be ineffectual, illegitimate, and have a difficult time maintaining itself. One can find instances in all governments where what limited police powers available are squandered, and where laws are flouted or circumvented. It is even more the case for taxing powers, where estimates are that as many as half the population believes it is legitimate to cheat if they can do so.[2] This is the case in the United States; and it is higher in many other nations. Poor design of government administration has the effect of undermining the legitimacy of public authority and is costly in every sense of the word. Authors David Osborne and Ted Gabler have such concerns in mind when they exhort policy makers to employ measures that rest lightly on society, that do not require so much 'muscle,' what they call 'Catalytic Government: Steering Rather than Rowing.'[3] Skillful design husbands the resources of government.

What makes the challenges of public administration even more difficult is the realization that both tools are better at circumscribing, or even stifling, behavior it opposes rather than promoting it. Bear in mind that any public policies typically have costs – either in the public resources required to administer them or by reducing general welfare. Care must therefore be taken to ensure that their design should be considered and explained. Examples abound where policies, typically with commendable goals, have been implemented, but with consequences that are unanticipated and often harmful and expensive. Often too it is the symptoms of problems that are addressed rather than the underlying causes, the result being that they momentarily or provisionally supply answers but which further exacerbate situations in due course.

With respect to the two problems at hand, traffic congestion and sprawl, most governments have failed to adequately deal with their challenges because they have taken little pain to fully understand their genesis and root causes. With respect to traffic, for example, the solution has too often been to build more roads, or else to widen them. Yet it has long been understood that such policies usually foster greater traffic congestion. Among students of systems theory this has come to be known as Braess's Paradox, after the brilliant German mathematician who first explicated it.[4] So it eventuates that most city thoroughfares of the world are plagued with overuse that is the direct consequence of foolish and counterproductive public policies. Other illustrations of transportation mismanagement could also be offered, but this example illustrates the point.

With respect to the matter of the centrifugal forces of sprawl development, it is more directly a result of economic misunderstanding. But the solutions seldom employ economics; rather policy makers look to command-and-control approaches like zoning and urban growth boundaries (UGBs). The earliest UGB was instituted decades ago in Portland, Oregon, advocated mostly by farmers whose land was threatened by the growing incursion of housing sprawl.[5] A girdle of protected green space was drawn around the city's perimeter, intended to prevent development in identified areas and presumably beyond it as well. In time, however, development leapfrogged the UGB and led to more commuter traffic and congestion beyond the girdle. Those property owners inside the perimeter were overjoyed with the arrangement because the scarcity enhanced their site values. Since locational values are a function of access and are reflected in capitalized transportation costs, people had the choice of paying more either in site rent for the privilege of location or else in travel costs from areas with more modest costs.

Ultimately the disequilibrium pressures of the site values inside and outside the UGB became so disparate that the system burst. Political forces reached a point wherein the differences could not be maintained and the UGB could not hold. Other cities have also attempted to delimit their suburban growth but have in one way or another faced similar problems. California's Bay Area outlined a growth boundary demarcated so far from urban cores that the projected infill would take a century. Political resistance made it impossible to impose it closer in where it would have greater bite. It encompasses an area larger than that of the US states of Connecticut and Rhode Island combined! So it is a meaningless pretense. Melbourne, Australia, has just reported a similar failure to curtail sprawl development, having relied upon a UGB pattern that has now been shown to encourage land speculation.[6]

Each of these examples reflects poorly conceived public policies, in the first instance an attempt to invest in greater infrastructure and essentially 'buy' a way out of the problem, and in the second case a misuse of a police-power-based command-and-control approach that simply postponed and amplified the problem. A better sense of economics, especially land economics, could have successfully addressed the challenge. Proper use of constitutionally permitted tax powers, the fiscal approach, can not only correct the distortions that result from market disequilibria; it can also raise revenue for the support of public services and obviate reliance upon revenue streams that have more negative impact and downside consequences. In exploring the problems at hand, the best solutions are in institution of a variant of the conventional property tax, what is most commonly known as land value taxation.

A BETTER SOLUTION IS A TAX ON LAND VALUES ALONE

The conventional real property tax as known in most English-speaking countries is really two separate taxes from an economic point of view: a tax on land values and a tax on improvement values. Each has very different dynamics and each influences behavior in a different way. Any tax on improvements, essentially buildings, penalizes upkeep and construction initiatives. Titleholders maintaining and improving property to the full extent that sites warrant are hit with a higher tax. Owners that let property go to wrack and ruin are rewarded with lower assessments and hence lower taxes. Just as taxing wages discourages work, as taxing interest discourages savings, and taxing sales discourages consumption, taxing improvements to real estate rewards the wrong behavior. There are long histories of tax folly from the time trees were taxed effecting deserts, when taxing windows led to darkness, and taxing lot frontage led to outlandish 'shotgun houses' in the old American West.

On the other hand, the tax on the assessed value of the land component of a parcel encourages investment and development. The higher the tax the more the owner is encouraged to build on the site so as to recover his carrying costs. Heavier taxing of underused and vacant parcels generates improvements, especially in high-value urban cores.[7] This development then fosters the necessary density to make localities walkable and less vehicle dependent. The vehicles that then service the areas tend to be public transit. The tax on land values and the tax on improvement values are like a train with an engine on each end: they work in opposite ways and negate what powerfully beneficial effects a tax on land value alone has.

Since the primary concern of this chapter is environmental policy and the use to which revenue streams can be put to accomplish sound environmental goals, I will return to this line of thought shortly. It is important, however, to recognize that a tax on land values comports perfectly with all the principles of sound tax theory. Among them are efficiency, neutrality, equity, administrability, stability, and simplicity. An ideal tax is neutral and efficient with respect to markets and progressive in so far as those who have fewer resources pay less. A soundly based land tax is also easily administered, simple to understand, and provides a stable and reliable revenue stream. It is certain in the face of any attempts at evasion. One cannot escape a land tax by taking it to the Isle of Man or the Cayman Islands. Many students hold the view that all taxes have downside attributes so that any revenue system must necessarily make compromises and trade-offs. This claim is very much open to challenge. It is important here only to emphasize that taxes impact behavior in ways that go far beyond their purposes of supporting public services. To this extent, their architecture needs to be carefully designed and understood.[8]

Tax principles as enumerated above have been recognized in various ways since first set forth by Adam Smith.[9] But in recent years there are considerations above and beyond those relevant to revenue design itself. Environmental concerns are equally important, particularly as they address land use configurations. Taxing land parcels according to their market value fosters land use patterns that best suit the demands of the community as a whole. Those sites that reflect where people want to be command the highest land values; those sites that are of marginal use or concern from a pricing perspective are taxed less and reflect less pressure to improve. Business and commercial parcels tend to cluster in high value areas, residential parcels develop at the edges, and agricultural and forest property is relieved of pressure to develop and consume land.

TAXATION OF NATURAL RESOURCE RENTS

The market value of land parcels reflects what classical economists called rent, also called ground rent or economic rent. Although originally thought of as applying to the productivity of farmland, rent is today identified far more with urban space. This is a different meaning of the word rent than when paying someone for the use of some property, whether for things like tools or for real estate purposes. Land rent is a flow of value through any natural resources that command a market price on account of their demand. Classically, rent is the market price any such commodity beyond whatever expense is needed to bring that factor into use. It applies as much to air or water or mineral and petroleum resources as it does to locations. Any resources that have a market price not created by human hands or minds can have rental value – even airport timeslots, electronic signals, and satellite orbits. Because their value results not from any human efforts, resource rents can be understood as socially created wealth; they are the mutual result of common enterprise and such rents flow through property more than they are generated by it.

Site rent can also be construed as capitalized transportation costs.[10] Sites with high market value are easily accessible; by whatever mobility means are at hand. Land sites reflect all such costs – those borne by individual members of society as well as those borne collectively. Site rents and transportation costs in a metropolitan area are essentially reciprocal: parcels in urban cores have high access and rental value whereas parcels in remote areas have high transportation costs and low rental value. One way or another the people have to pay for access to market exchanges, whatever sort or style they have: one pays either for the privilege of occupying a location or for the cost of getting there.[11] But since transportation costs reflect the use of materials and energy, it makes sense that they be efficiently consumed, important for a well-designed locality.

German economic geographer Heinrich von Thunen worked out this theory almost two centuries ago.[12] He calculated the costs of bringing farm goods to market and as they related to the most suitable distance from the market on which to grow them. The reciprocal of this was his understanding of the value of the market sites themselves. He appreciated that locations in urban cores had site prices many times those in agricultural areas. He further understood the relationship between site rents and access. Von Thunen died before the carbon age was fully upon us and before transportation costs became for the moment almost inconsequential. We live today in a time of temporary luxury when it comes to energy consumption, an age which most believe will soon pass.[13] Given how intractable land use configurations are once set in place, societies are foolish to develop a permanence that make them far less livable once the petroleum age largely passes.

Returning once more to the matter of the flow of ground rents through locations, one needs to understand that if the public does not recapture the socially created rent it is capitalized in lump-sum market prices. For titleholders to such sites this constitutes windfall gains, what John Stuart Mill called an 'unearned increment.' This is surplus wealth reflecting social productivity that becomes effectively frozen and unavailable as resource capital. Moreover, if this flow of ground rent is not taxed and restored to the economy, the public then is forced to rely on other taxes that have more downside impacts. As earlier noted taxes on wages and goods discourage economic vitality, distort market choices, and are administratively expensive to collect and enforce. Lastly, when entrepreneurs or households make real estate investments they are usually forced to pay artificially inflated prices for locations where value is fed by speculative practices. Members of society pay twice as a result, first for real estate investment loans and then again in taxes to support public services. The only winners are speculators and bankers.

When speculators keeping them off the market inflate land prices, those who would elect to use those sites were they available are forced to choose second-best and sub-optimal locations instead. Rather than market-clearing efficiencies assuring the rational development of social spaces, one finds leapfrog and haphazard unfolding settlement. All this adds to extra costs in infrastructure – roads, utility services, public amenities and community services – that are also less than optimal in their provision. Spatial arrangements thereby impose their social costs several times over, all of which lead to community well being that is far below what could be optimally obtained.

Chances for private capture of socially created rental value of land arose only during the past four centuries of the 'great land grab.' From roughly 1650 on, natural resources that earlier were regarded as part of the public commons were turned into a marketable commodity and privatized for selfish gain.[14] This was rationalized and justified in numerous arguments and judicial decisions.[15] The world is only now beginning to appreciate the implications of this rush to privatization. One could argue from an economic perspective that the leasehold arrangements that characterized classic civilizations were equal to or better equilibrated than are the tax regimes employed in nations today. The excess use of natural resources resulting from treating them either as 'free goods' or commodities captured by whatever parties secured legal titles has meant the impoverishment of everyone, and even jeopardizes the sustainability of the earth.

This 'enclosure movement' was initiated during the Tudor era of English history[16] and it would be difficult now to recapture and restore much of this property to the public realm. A more promising solution is to collect the rent from land parcels based upon their market price, treating land not as a commodity to be owned by title in fee-simple but rather as a usufruct. This policy not only encourages the economy to perform far more efficiently, it also restores a sound moral basis to the economy and offers a clear theory of distributive justice. That which is rightfully the public's is returned to the public; that which is created by one's own mind or body is one's own to possess.[17]

The concept of usufruct ownership, in contrast to fee-simple title, is a term that needs to be restored to contemporary discourse.[18] It constitutes the legal right to use and benefit from property, typically natural resource property, that other persons, institutions, or the general public have formal title to, at least so long as the property is not damaged or degraded. The English word usufruct derives from the Latin expression *usus et fructus*, meaning 'use and enjoyment,' cognate to English 'use and fruits.' The concept of usufruct goes back to ancient times, and has been far more evident in societies of the world than the notion that elements of nature can be owned as commodities. Thomas Jefferson wrote, citing John Locke, that 'the land belongs in usufruct to the living,'[19] a quote that Henry George often repeated, as in his noted speech, 'The Crime of Poverty.' George also held that private capture of that which was God-given constituted theft, pure and simple: 'Thou Shalt Not Steal!' he told the Anti-Poverty Society of New York in 1887.[20] Native American people put the same principle differently: we do not inherit the earth from our ancestors; we borrow it from our children. Even in American society where private property in land and nature is a sacrosanct hallmark of its version of capitalism, the law does not talk about it in such terms. Rather it talks about property ownership as a 'bundle of rights.'[21]

Prior to the enclosure movement, the use of land was typically paid for in various forms of rent. Once can trace the origins of such payments to earliest times and show that such practices were almost universal. Historically rent payments were usually a part of a farmer's yield or a specified number of days of corvée labor. Rent payments were made to society, or to nobility acting in its name, often also in the form of tribute goods.[22] Based on existing records, it appears that rent surplus was usually about a third of a society's economy, more on richer land less on poorer.[23] An old English nursery rhyme reflects this common practice in feudal arrangements:

Bah, Bah black Sheep, Have you any Wool? Yes Sir, Yes Sir, Three Bags full. One for my Master, One for my Dame, One for the little Boy That lives down the lane.[24]

LAND RENT IN A MODERN ECONOMY

There is growing appreciation among some economists, urbanologists, tax theorists, and land use planners that the disregard or trivialization of land, and its associated rent, as a factor of production has had profound consequences for many aspects of social and economic evolution. Failure to recognize the significance of the flow of rents from land and other natural resources has led to distortions in many realms of society and their economies. It was possible to overlook this distortion so long as there lacked the means by which to identify and quantify it; rent was posited and discussed largely in economic theory, and what means were available to identify and quantify it were for the most part derivative and inferential. Computer power and the availability of quantifiable data now offer greater opportunity to redress this failing, and evidence of its importance and impact mounts.

Capturing socially created resource rents restores to liquidity elements of the active economy that are otherwise 'frozen capital;' this improves market efficiency and productivity. We know also that absent taxes on labor and goods the amount of rent would be much greater. After all the shifts in their incidence through the economy, 'all taxes ultimately come out of rent.' This is an axiom that has come to be known by the acronym ATCOR.[25] Sometimes it is framed that 'all taxes are at the *expense* of rent.' Put still another way, total rent is that remaining *net* of taxes.

One quick study estimated that a full land tax, excluding rents from pollution rights, the electromagnetic spectrum, landing slots, corporate charters, internet addresses, and other sources, would yield rent amounting to about 28 percent of GDP,[26] and a far more detailed and sophisticated study of the total land rent in Australia estimates that the total is well above thirty percent of GDP. It concluded that, 'the "bottom line" reinforces the overall conclusion … that land-based tax revenues are indeed sufficient to allow total abolition of company and personal income tax.'[27] A full enumeration of sources where additional rents situate would take enormous work, but Mason Gaffney has suggested 15 major places as a start, all of which by their private capture now reduce economic productivity.[28] When all is said and done, he suggests, 'The Hidden Taxable Capacity of Land [is] Enough and to Spare' and could support government and supplant all present taxes.[29] This finding corroborates George's original argument.[30]

> In every civilized country, even the newest, the value of the land [i.e. the amount of taxable rent] taken as a whole is sufficient to bear the entire expenses of government. In the better-developed countries it is much more than sufficient. Hence it will not be enough merely to place all taxes upon the value of land. It will be necessary, where rent exceeds the present government revenues, commensurately to increase the amount demanded in taxation, and to continue this increase as society progresses and rent advances.

In the past 30 years, economists of major stature have demonstrated the validity of this claim, now known as the Henry George Theorem.[31] Gilbert Tucker, a self-taught student of Henry George, also anticipated this in a short book titled *The Self-Supporting City*. In it, he begins boldly by arguing,

> Municipal taxation as now levied can and should be a thing of the past: the American city can be a self-supporting corporation, meeting its expenses from its rightful income. Taxation is unnecessary, because the city has, in its physical properties, acquired through the years, by the expenditure of its people's moneys, a huge capital investment from which it collects only a very small part of the return earned.[32]

The virtue of taxing rent is that it captures unearned income that is otherwise windfall gains to households and businesses. It is typically the wealthier elements of the population that have title to property resources, so that to them the capture of untaxed but socially created rents constitutes a 'free lunch.'[33] The component of the population that owns no land of any sort, typically the poorest elements of society, pay no rent taxes at all for the reason that resource rents, coming from sources with inelastic supply, cannot be passed forward. This makes the taxation of land rents highly progressive, besides their possessing all the other attributes of a sound tax structure.[34]

Most of the literature exploring the nature and sources of economic rent tends to focus on what flows through surface locations of the earth, what is known as ground rent or land rent. Very little attention has been given to rent sources from other elements of nature, since ground rent now seems to be the largest single component. But a significant additional source of resource rent is generated from minerals and fossil fuel extraction. Consider also the wealth of the world's oceans, mostly used today as a source of fish. The aforementioned spectrum, the frequencies on which radio, television, mobile phone and other signals travel in today's world, all yield economic rents. And especially now, the air sink itself must be recognized for its rental value to the extent that it is used for pollution emissions, and to the extent that it is capable of absorbing them.[35] The air, after all, is rightfully the birthright of all humanity, and its sale to polluters to use as a dump is the penultimate travesty in the privatization of the commons.

CONCLUSION

It is now clear that identifying and recapturing the full measure of land rents would make for sounder tax policy, restore it to a moral framework, facilitate better use of natural resources, and make for greater livability and sustainability. Even more importantly, it would restore to all people a commons that is rightfully their birthright.

NOTES

1.   There are also, of course, war powers, peripheral to domestic policies and not considered further here.
2.   Bartlett, Donald L. and James B. Steele (2000), *The Great American Tax Dodge*. Boston: Little, Brown & Co.
3.   Osborne, David and Ted Gabler (1993), *Reinventing Government*. New York: Addison Wesley.
4.   Steinberg, R. and W.I. Zangwill (1983), 'The Prevalence of Braess's Paradox.' *Transportation Science*, Vol. 17, no. 3, 301–18.
5.   Batt, H. William (2003), 'Stemming Sprawl: The Fiscal Approach,' Chapter 10 in Lindstrom, Matthew J. and Hugh Bartling (eds), *Suburban Sprawl: Culture, Theory, and Politics*. Lanham, MA: Rowman & Littlefield Publishers, Inc. and www.cooperativeindividualism.org/batt-h-william_stemming_sprawl.html, accessed January, 2011.
6.   Johanson, Simon (2010), 'Land Prices Shatter Mortgage Belt Dreams,' *The Age* (The Age.com.au), December 6.
7.   A 1995 study of Pennsylvania cities using such a tax concluded: 'The results say that in all four categories of construction, an increase in the effective tax differential [between land and buildings] (1) is associated with an increase in the average value per permit. (2) In the case of residential housing, a 1% increase in the effective tax differential is associated with a 12% increase in the average value per unit ... From the perspective of economic theory, it is not at all surprising that when taxes are taken off of buildings, people build more valuable buildings. But it is nice to see the numbers.' Tideman, Nicolaus and Florenz Plassmann (2000), 'A Markov Chain Monte Carlo Analysis of the Effect of Two-Rate Property Taxes on Construction,.' *Journal of Urban Economics*, 47 (2), 216–47.
8.   See www.progress.org/cg/battprincip02.htm. accessed, January, 2011.
       See also Batt, H. William (2005), 'The Fallacy of the "Three-Legged Stool" Metaphor' *State Tax Notes*, 35 (6), 377–81, www.cooperativeindividualism.org/batt_on-tax-policy.html accessed, January, 2011.
9.   In Smith, Adam ([1776] 1937), *The Wealth of Nations* (New York: Modern Library edition, p. 796), he concluded, 'Ground-rents and the ordinary rent of land are ... the species of revenue which can best bear to have a peculiar tax imposed on them.'
10.   Batt, H. William (2008), 'Modeling Land Rent and Transportation Costs in the United States,' in Janet Milne, Kurt Deketelaere, Larry Kreiser, *et al.* (eds.), *Critical Issues in Environmental Taxation: International and Comparative Perspectives: Volume 1*. London: Richmond Law & Tax, Ltd.

11. To be sure, individuals and organizations seldom experience all these costs because many of them are socialized. One study showed that the average motor vehicle owner only pays 10 percent of the true costs of his driving. Miller, Peter and John Moffet, (1993), *The Price of Mobility: Uncovering the Hidden Costs of Transportation*. Washington, D.C.: Natural Resources Defense Council, and MacKenzie, James J. (1992), *The Going Rate: What It Realty Costs to Drive*. Washington, D.C.: World Resources Institute.

12. Thünen, Johann Heinrich von (1826), *The Isolated State*. Oxford: Pergammon Press (1966).

13. See, for example, the writings of Heinberg, Richard (2003, 2004, 2008, 2010), *The Party's Over, Power Down, The Post Carbon Reader*, and *Peak Everything*, all from Gabriola Island, BC, Canada: New Society Publishers, and www.hubbertpeak.com/, accessed January, 2011.

14. Chandler, Alfred N. (1945), *Land Title Origins: A Take of Force and Fraud*. New York: Schalkenbach Foundation, and Weaver, John C. (2003), *The Great Land Rush and the Making of the Modern World, 1650–1900*. Montreal: McGill-Queen's University Press, and many others.

15. The legitimacy of titles to real property has recently been described and explored in what is commonly called the 'Doctrine of Discovery.' See Miller, Robert J. (2006), *Native America, Discovered and Conquered: Thomas Jefferson, Lewis & Clark, and Manifest Destiny*. Westport, CT: Praeger. Stated another way, it is 'Finders, Keepers.' My review, listing principles of the Doctrine of Discovery, is online at www.cooperativeindividualism.org/batt-h-william_review-of-native-america.html. accessed, January, 2011.

16. Polanyi, Karl (1944, 1957), *The Great Transformation: The Political and Economic Origins of Our Time*. New York: Reinhart & Co., and Beacon Press.

17. By the same token, taxing the product of one's labor or of the labor itself is more ethically questionable. In the case of privately captured wealth from economic rent, one is hard-put to justify it at all. John Stuart Mill, an early proponent of public capture of rents, argued 'landlords grow richer in their sleep without working, risking or economizing. The increase in the value of land, arising as it does from the efforts of an entire community, should belong to the community and not to the individual who might hold title.' Mill*, John Stuart (1994), *Principles of Political Economy*. London: Oxford University Press World Classics, Bk. 5, chap. 2, sec. 5. Supportive quotes similar to this are easily found for Blackstone, Jefferson, Paine, and many other leading figures of the 18th and 19th centuries. It was Henry George, the last great moral integrator and defender of classical economic theory, who argued this in his monumental 1879 work, *Progress and Poverty: An Inquiry Into The Cause of Industrial Depressions and of Increase of Want with Increase of Wealth … The Remedy*. New York: Robert Schalkenbach Foundation.

18. A collection of quotations and uses are found at www.wealthandwant.com/themes/ Usufruct.html accessed, January, 2011.

19. Jefferson letter to James Madison, Paris, Sep. 6, 1789.

20. Henry George, 'Thou Shalt Not Steal,' an Address delivered on May 8, 1887 to the Anti-Poverty Society, New York City. Henry George argued that seizing private titles to land, and by extension any elements of nature, was essentially theft, and was the moral equivalent to owning slaves. See www.wealthandwant.com/HG/George_TSNS.html accessed, January, 2011.

21. See, for example, Friedman, Jack P., Harris, Jack C., Lindeman, Bruce, *et al.* (eds). (2004), Barron's Educational Series: *Dictionary of Real Estate Terms, Sixth Edition*, p. 69; also at www.answers.com.

22. See especially, Lenski, Gerhard (1966), *Power and Privilege: A Theory of Social Stratification*. New York: McGraw Hill, 1966, especially Ch 9, and Fetter, Frank A. 'Rent,' *Encyclopedia of Social Sciences*. New York: MacMillan (First Edition, 1934), Volume XIII, pp. 289–92.

23.   Bloch, Marc (1966, 1970), *French Rural History: An Essay on Its Basic Characteristics*. Berkeley: University of California Press. p. 72; Bennett, H.S. (1937, 1971), *Life on the English Manor: A Study of Peasant Conditions, 1150–1400*. Cambridge: Cambridge University Press. Ch V: 'Rents and Services,' pp. 97–125, *passim*; and Bairoch, Paul (1991), *Cities and Economic Development: From the Dawn of History to the Present*. Chicago: University of Chicago Press. p. 283. Gerhard Lenski (*supra* note 22, p. 226) notes, however, that the Chinese Gentry was able at times to collect as much as 40–50 percent of the land yield as rent.

24.   This rhyme is also traceable to France as far back as the 17th century. See Wikipedia and other sites.

25.   For several further discussions of ATCOR, see www.wealthandwant.com/themes/ATCOR.html, accessed January, 2011.

26.   Cord, Steven (1985), 'How Much Revenue Would a Full Land Value Tax Yield?,' *American Journal of Economics and Sociology*, 44(3).

27.   Dwyer, Terry (2003),'The Taxable Capacity of Australian Land and Resources,' *Australian Tax Forum*, January. Governments in developed societies usually comprise from about 25 to 30 percent of GDP. It is difficult to generalize because many functions that are privatized in one nation are public in another. See http://carriedaway.blogs.com/carried_away/2003/10/us_government_s.html, accessed January, 2011. The US Congressional Budget Office calculates that Federal Government Outlays from 1962 to 2001 range from roughly18 to 20 percent. See http://carriedaway.blogs.com/carried_away/2003/10/us_government_s.html State and local governments constitute the remainder, accessed January, 2011.

28.   Gaffney, Mason (2004), 'Sounding the Revenue Potential of Land: Fifteen Submerged Elements,' *Groundswell,* Sept.-Oct., and www.progress.org/cg/gaff1004.htm, accessed January, 2011.

29.   Gaffney, Mason (2009), 'The Hidden Taxable Capacity of Land: Enough and to Spare,' *International Journal of Social Economics*, 36, (4), 328–411.

30.   George. *Progress and Poverty*, *supra* note 17.

31.   Especially notable are Nobel Laureates William Vickrey and Joseph Stiglitz. Professor Stiglitz's most recent paper is 'Principles and Guidelines for Deficit Reduction,' Roosevelt Institute December, 2, 2010, www.rooseveltinstitute.org/people/fellows/joseph-stiglitz, accessed January, 2011. For a more complete list and discussion of this argument, including an account by Gilbert Tucker (1946, 1958, 2010) anticipated all this in his *The Self-Supporting City*. New York: Robert Schalkenbach Foundation. The 2010 edition has an afterward by this writer.

32.   Tucker, *The Self-Supporting City*, *ibid*. p. 1.

33.   This belies the declaration by neoclassical economist Milton Friedman (1977) that there is *There is No Such Thing as a Free Lunch*. New York: Open Court Publishing.

34.   Numerous materials are available to attest to this. Among the most accessible websites, with links to many others, are www.cgocouncil.org, www.theIU.org, www.Schalkenbach.org, www.urbantools.org, www.labourland.org, www.cooperativeindividualism.org/, www.earthrights.net, www.prosper.org.au, accessed January, 2011.

35.   See Collier, Paul (2010), *The Plundered Planet: Why We Must – and How We Can – Manage Nature for Global Posterity*. New York: Oxford University Press; Gaffney, 'Fifteen Submerged Elements,' *supra* note 28, and www.progress.org/cg/gaff1004.htm and Gaffney, 'Hidden Taxable Capacity,' *supra* note 29, accessed January, 2011.

8. Land management and local taxation in Italy

Giorgio Panella, Andrea Zatti and Fiorenza Carraro

1. INTRODUCTION

The phenomenon of land consumption by urbanization has been growing increasingly in the last decades. Urbanization competes with agriculture for land use and prompts for the occupation of marginal lands, sometimes even those unsuitable for settlement.[1] This trend is spread all over several countries and poses serious problems (EEA, 2006). Soil depletion, environmental externalities and the provision of local public goods have pushed public authorities to intervene through different distributive instruments.

The traditional solution to these kinds of problems has always been found in making local communities pay the costs involved in the provision of public goods through taxes (Brueckner, 1997). Most recent directions suggest linking the tax to the matter under discussion: the environmental externality or the urbanization costs. Therefore, traditional instruments such as zoning or urban planning have been coped with by using economic measures such as urbanization charges or construction fees.

The coexistence of instruments and measures different in nature and function has sometimes raised the question of policy overlapping. In fact, it has been argued whether or not the use of quantitative regulations (zoning), on the one side, and the employment of fiscal instruments, on the other, creates an excessive regulation, with the result of neutralizing the beneficial effect of the first kind of measures. To date, national and international experiences support the idea that the two kinds of instruments can jointly function properly, given that they pursue different aims. Zoning normally represents the most efficient way to determine the optimal quantity of building area; fiscal instruments, instead, are suited to deal with negative externalities and the provision of public goods.

This chapter deals with the use of fiscal policy at the local level and the proper definition of urbanization charges/duties to promote a sustainable consumption of lands.

After a brief introduction on the main challenges posed by land management (Section 1), Section 2 analyses the theoretical aspects underpinning local policies in urban planning. Section 3 analyses the way land management is dealt with in Italy, focusing on the use of construction charges. Section 4 deeply evaluates the use that has been made of construction charges in the territories of Italy, paying particular attention to the pitfalls and negative outcomes incurred: the predominance of budget needs with respect to balanced urban development (Section 4.1); the fiscal competition between local municipalities (Section 4.2); municipal debt (Section 4.3); the unsustainable consumption of land (Section 4.4). Section 5 concludes with suggesting policy proposals on finance and governance approaches that could improve the environmental sustainability of land and the economic sustainability of local budgets.

2. LAND MARKET AND PUBLIC INTERVENTION

Public authority can rely on a wide variety of instruments and measures to deal with land management and urban planning. Normally, the actual land market is not purely competitive[2] and several kind of negative externalities are involved (Arnott, 1987). The supply or transfer of urbanized lands generates, along with private marginal costs, two different types of marginal social cost: the negative externalities, such as the deterioration of environment and landscapes, and the marginal damage caused by the failed production of public goods (the loss in social benefits coming from the underproduction of public goods).[3] Since external factors are not perceived by private operators, it is necessary to intervene in order to bring back the land market to efficiency criteria. The quantity of land suitable for building should be identified where private marginal benefits equal social marginal costs, made up of private costs and social damages (Malpezzi, 1999).

The most common intervention in this field consists of fixing a cap for the amount of land suitable for building (zoning). This method would reduce housing density (which private operators would like to increase) and would preserve spaces for the provision of public goods (which private operators would underproduce). Zoning establishes land uses and the ways in which property can be exploited. By attributing specific designations to land, it attempts to avoid incompatible uses that can be mutually damaging (Micelli, 2002). The method of zoning is a convenient instrument for urban planning as a whole, given that it helps to identify the lands suitable for urbanization, those devoted to agriculture and finally those left to public use. However, this situation creates equity problems. Preventing some lands from being urbanized, results in land owners experiencing an economic loss. The public authority can compensate their loss with monetary transfers. However, these compensations, together with the expenditures related to the provision of public goods, can make the public budget unsustainable.

The predominance of urban planning has been in most cases associated with the revulsion of any kind of price driven method. Urban planning and zoning work basically on the decision to set a given quantity of land to be urbanized; and this decision can even largely diverge from what market forces would identify in terms of efficient allocation of lands. In order to avoid some of the equity and distributive problems of zoning, many other instruments, mainly based on economic incentives, have been developed.

A major step has been the development of market-based measures, such as development rights or fiscal instruments, potentially able to restore conditions of efficient resource allocation.

Italy applies a mixed system comprehensive of urbanization charges and planning permissions. These measures have shown strengths and pitfalls and much can be done to encourage the sustainable management of land.

3.   CONSTRUCTION CHARGES

Law n. 10/1977 (Law Bucalossi) established that every activity entailing the urbanization of a territory within the municipality shares the expenses involved in land transformation and is subject to a building permission released by the mayor (Art.1). The release of a planning permission involves the payment of a financial contribution, the so called construction charge, commensurate with the expenditures of urbanization and the construction cost (D.P.R.[4] n. 380/2001).

The value of the construction charge, established by regional law, is therefore linked to two parameters:

● urbanization costs
● construction costs.

The first component can be traced back to an earmarked tax. That is, the private operator pays the public administration back for the provision of public goods such as urbanization works (roads, energy and water networks etc.). In this case the charge is interpreted as a one-off payment due in the case that the building activity entails an increase in the urban load.[5] The second component, instead, represents a sort of property tax. Construction charges are used to recover part of the revenues derived from the building intervention. The building contributions are in fact somehow linked to the construction costs and therefore to the building's market value.[6]

The incidence of the urbanization charges[7] is established, upon the decision of the city council, on the basis of parametric tables that the Regions define for different municipalities according to:

A.   the extent and trend of the population
B.   the geographic characteristics
C.   the zoned areas as established by the town and urban planning
D.   the limits and mandatory terms set by the Central State and Regions.

Concerning the construction charge, what matters is the future use of the building under construction. Three different categories of activities with respect to the building constructions are distinguished:

1.   new residential buildings
2.   retrofitting of existing buildings
3.   retrofitting of buildings not belonging to the residential sector.

Basically, the starting point for the charge setting is the definition of the fixed average costs of urbanization,[8] set differently according to the use that can be made of a certain building. Then, two coefficients are applied: one addresses the kind of intervention that has to be done (categories 1, 2 and 3); the other considers the specific features of the territory (categories A, B, C and D). Finally, these values are updated yearly with the ISTAT (National Institute of Statistics) revaluation coefficient.[9]

The construction charges are fiscal instruments of zoning that can be profitably used in conjunction with urban planning. They are meant to be conceived as 'special contributions', duties levied on those who gain an individual advantage from public activity; they have also been interpreted as a sort of environmental tax on the land use.

The payment of the construction charge should fulfil two specific requirements:

●   the expenditure coverage for the infrastructures linked to the new buildings (roads, energy and water networks, parking etc.)
●   the territory governance, guaranteeing that choices are made for the common good.

However, recently this tax revenue has been unbounded from its original scope. Nowadays, municipalities can use up to two-thirds of the revenue from construction charges to finance current expenditures of various types

(Cipollina, 2008). And this trend is expected to grow more and more now that the economic crisis obliges local governments to face urgent budget needs.

4. URBAN TAXES AND TERRITORY GOVERNMENT IN ITALY

In Italy, local municipalities rely on different funding sources when involved in the government of territories. Apart from the money transfers used to equalize the financial imbalances between local governments, the tax base is mostly made up of the existing building stock and the urban and productive development allowed by urban and territorial planning and regulations.[10] Basically, the economic and physical features of a territory affect the local budget management.

According to Law n. 10/1977, the main aim of the construction charge should have been to deter the extensive exploitation of urban land (Magnani, Muraro, 1978; Osculati, Zatti, 2010). However, this provision does not seem to have provided the expected effects. The construction charge represents one of the main funding sources for local municipalities and local revenues strongly depend on the existing building stock and on its evolution over time. Over the years, this fact has favoured urban development and sprawl (Ferri, 2007). The contraction in money transfers from the central government, on the one side, and the increased competences assigned to local municipalities, on the other, have pushed local governments to enlarge the tax base with urban development in order to keep the tax rates low (Curti, 1999). Moreover, the recent budget problems[11] faced by local municipalities made them devote the collected revenues to the general budget;[12] when previously they had been employed in the field of urbanization works.

Tables 8.1 and 8.2 provide a useful overview of the role played by the construction charges as revenue sources. In the period 2001–06, the collected revenues (as a percentage of total revenues) increased, supporting the idea that local governments have made extensive use of these instruments. However, their role in the coverage of capital expenditures decreased, given the fact that the D.P.R. n. 380/2001 allowed up to 50 per cent of the revenues coming from the construction charges to be devoted to the coverage of current expenditures and another 25 per cent to the asset maintenance.

Table 8.1 Construction charges – Italian regions 2001–06 (% of total revenues)

	2001	2002	2003	2004	2005	2006
Piemonte	3.8	3.9	3.6	4.1	4.3	4.8
Valle d'Aosta	1.8	2.3	1.7	2.7	3.1	3.1
Lombardia	4.1	4.4	5.2	5.7	3.5	4.6
Trentino Alto Adige	3.2	3.2	2.5	3.2	2.4	3.4
Veneto	5.0	5.9	5.9	6.6	5.3	5.1
Friuli Venezia Giulia	1.8	2.4	2.2	2.2	1.5	1.6
Liguria	1.9	2.2	2.8	3.2	3.7	3.0
Emilia Romagna	7.2	8.1	5.2	8.1	7.7	7.7
Toscana	4.8	6.6	7.0	8.3	6.9	7.5
Umbria	2.3	2.7	2.8	2.8	3.1	2.8
Marche	3.5	5.5	4.4	6.5	6.3	5.5
Lazio	3.0	4.3	4.4	9.0	6.5	5.3
Abruzzo	4.1	4.4	4.2	4.9	4.3	5.3
Molise	2.4	3.2	2.8	2.8	2.5	2.3
Campania	1.5	1.8	2.2	3.5	3.7	2.6
Puglia	4.1	4.1	5.5	5.3	5.7	3.9
Basilicata	2.6	0.0	3.2	2.0	2.7	2.2
Calabria	1.8	2.1	2.3	3.0	3.3	3.0
Sicilia	3.1	1.7	2.4	3.8	3.0	2.5
Sardegna	2.2	2.9	2.9	3.1	2.1	2.1
ITALIA	**3.7**	**4.1**	**4.3**	**5.5**	**4.4**	**4.4**

Source: IRPET on ISTAT and ISAE data.

*Table 8.2 Capital expenditure coverage through construction charges –
Italian regions 2001–06 (% and average values)*

	2004	2005	2006	Rate 2005–06	Average value
Piemonte	11.0	10.7	12.2	14	11.3
Valle d'Aosta	5.9	7.3	7.6	4	6.9
Lombardia	10.2	5.6	7.9	41	7.9
Trentino Alto Adige	6.4	4.9	7.8	59	6.4
Veneto	17.0	14.3	13.7	-4	15.0
Friuli Venezia Giulia	5.6	3.2	3.7	16	4.2
Liguria	9.7	13.0	11.2	-14	11.3
Emilia Romagna	22.8	23.7	23.2	-2	23.2
Toscana	25.4	24.3	24.4	0	24.7
Umbria	5.2	5.9	5.7	-3	5.6
Marche	15.5	17.0	15.3	-10	15.9
Lazio	34.5	22.0	16.6	-25	24.4
Abruzzo	11.3	10.4	14.5	39	12.1
Molise	7.7	7.5	6.0	-20	7.1
Campania	10.3	11.9	8.2	-31	10.1
Puglia	14.2	22.3	14.8	-34	17.1
Basilicata	5.2	6.8	5.8	-15	5.9
Calabria	10.3	11.8	11.7	-1	11.3
Sicilia	21.0	16.5	12.7	-23	16.7
Sardegna	7.7	5.1	5.0	-2	5.9
ITALIA	**13.8**	**10.9**	**11.3**	4	12.0

Source: IRPET on ISAE data.

The main negative outcomes of the fiscal autonomy left to the local
municipalities in the field of urbanization have been:

● the general idea that budget needs come first with respect to urban
  sustainability; with the consequence of rapid loss of agricultural land
  to urbanization
● fiscal competition between local municipalities and lack of a common strategy

- urban expansion, increase in the costs of public services and, therefore, municipal debt
- unsuccessful assessment of the non-monetary values of land
- extra rents on behalf of certain categories of persons at the expense of others.

It is difficult to find a solution to these problems. In most cases they are caused by exogenous factors such as the evolution of lifestyles, the fall of transport costs, the contraction of journey times and the lengthening of distances (Magnani, 2001). Given that land prices do not correctly reflect the different advantages of localization, it is necessary to think about new instruments able to represent the value enclosed in each plot of land. In this field, local taxes, tariffs and urbanization duties could be properly defined so as to internalize the effects of urban sprawl.

4.1. Construction Charges, ICI and the Building Cycle

Municipal budget revenues strongly depend on the building cycle. It seems there is a certain degree of correlation between urban planning and budget revenues.

The existence of many industries and services over a certain urban area guarantees more consistent revenues for the local budget, either in per-capita terms or as a percentage of total municipal revenues (ISPRA, 2008). This is especially the case for revenues coming from construction charges. The local council property tax (*Imposta Comunale sugli Immobili, ICI*), instead, being a sort of property tax, seems not to be related to the productive structure of the territory.

The general purpose of both the ICI and construction charges is the same: catching part of the building rent. However, they differ in the way the revenues are collected. The tax base for the ICI is the building cadastral rent; the construction charges are somehow related to the construction costs. The ICI has the primary aim of catching part of the private rent established by the building market and, notably, it is due every year. Construction charges, instead, must be paid only once, at the beginning of construction work and they are due to get partial coverage of the expenditures involved in the provision of local public goods (Agnoletti, Ferraina, 2010).

Building activities are more and more subject to the price fluctuations of building and capital markets; this circumstance brings about the exposure of municipal budgets. In fact, if building activity is in a declining phase, it is not clear who would cover the expenditures related to the development of public urban infrastructures, given that the revenues coming from the construction charges decrease. On the contrary, urban development enlarges the tax base for ICI and construction charges; while at the same time it feeds the demand for urban infrastructures and services[13] (Curti, 2004).

In this situation, it becomes necessary to guarantee a certain degree of flexibility to the instruments used for urban planning. Relying on a single measure seems not to be a winning strategy. Rather, it would be optimal to use a policy mix where the instruments included in the package would basically be commensurate with each specific circumstance according to the functioning of the building market, favoring the desired distributive impacts.

4.2. Territorial Competition

A major problem caused by the fiscal independence gained by local municipalities is territorial competition. It involves resources and land depletion and an unequal distribution of private and public activities.

When the functioning of a local municipality strongly depends on the revenues collected at a local level, there is an incentive to enlarge the tax base. This is what has happened in Italy over the years, when certain local municipalities have tried to increase their expenditure capacity by enlarging their tax base through urban sprawl, at the expense of the neighboring towns (Ave, 1999).

In the case of metropolitan areas, where each inner urban reality is ruled in a different manner, fiscal competition can cause several pitfalls. Municipalities with low local tax rates benefit from the settlements of productive activities; while neighboring municipalities do not. In order to limit these effects and contain or at least lessen the distortions they bring about, it is necessary to reduce the competition among different local tax rates and tax bases. First of all, metropolitan areas should be ruled by metropolitan regulations and prescriptions. The ultimate goal is to turn the horizontal fiscal competition between territories into a cooperative game, keeping in mind that human resources are mobile while natural resources are not (Ferri, 2007).

This idea comes from the international experience of 'tax base sharing'. The purpose is to create a compensation fund[14] so as to reduce the difference of revenues existing among territories suitable for building and areas subject to agricultural or environmental constraints (Ceriani *et al.*, 2008). In the case of metropolitan areas, the tax base sharing method asks each local entity to provide the metropolitan fund with a certain percentage of their revenues. The collected resources would be subsequently redistributed among the local entities in accordance with their geographic dimension and budget needs. The compensation fund, therefore, is expected to contain the unbalanced distribution of negative and positive externalities among the territories. Last but not least, tax base sharing enables the accomplishment of economies of scale and scope, with a probable consistent money saving.

4.3. Municipal Debt

Urban expansion and the increase in population pose serious problems in terms of municipal debt. The next generation is expected to face not just the environmental damages caused in the present, but also the financial risks related to the current municipal debt.

Since Law n. 10/1977, the main aim of the construction charges has been to cover part of the expenditures involved in the realization of urbanization works, such as energy and water networks, roads etc. However, recently some normative regulations (D.P.R. n. 380/2001, Budget Law 2008) have lessened the role of these contributions in the funding of urban planning. Basically, local governments rely on construction charges to cover part of their municipal debt, in such a way generating a vicious circle. More and more planning permissions are released to get revenues for the local budget's current expenditures and less is left for the financing of public infrastructures. Who is going to pay for that? How can we solve the equity and distributive problems that might arise?

According to some recent theories, the tax rates related to construction costs should be differentiated between new and old residents (Gilbert, Guengant, 1993).[15] Others suggest increasing the tax rates of the so called 'entrance taxes': urbanization charges, construction charges, and ICI on building areas. Their increase would contribute to the maintenance costs of at least the existing infrastructures. Moreover, an increase in the tax rate of the entrance taxes, equally shared by a certain group of municipalities, would help countervail equity and localization problems prompting more sustainable land depletion.

4.4. Land Consumption: What Can be Done?

The link between population increase and economic growth on the one side and urban development on the other is no longer linear: urbanization is more and more pervasive and complex. In the last decades, it has experienced an unprecedented acceleration, basically independent from any demographic and economic trend. This phenomenon, spread all over the world, is particularly worrying in Italy, where land scarcity has prompted for the occupation of even marginal lands (ISTAT, 2009).[16]

In order to collect the necessary revenues to cover urbanization expenditures and contemporaneously discourage the excessive consumption of land, local governments have to modify fiscal and planning instruments so as to reflect the loss of land to urbanization. The proper definition and differentiation of tax rates, along with the use of fiscal incentives and disincentives, can in fact modify the consumers' preferences in the short term as well as in the long term. The reasoning underpinning this mechanism is to provide private investors with certain market signals so as to modify their economic interests. These signals are meant to be given by taxing the land use, given that the land exploitation for certain purposes can generate negative externalities on the environment and the society.

Some of the measures that can be implemented with the aim of preserving agricultural lands are:

- modulation of the urbanization and construction charges and of any kind of fiscal incentive according to the land use and the land loss[17]
- tax relief for the building activities which do not involve an increase in the urban load
- an ICI tax cut (or exemption) for high quality agricultural lands and landscapes.

In order to reduce land depletion and at the same time guarantee a sufficient level of budget revenues, the construction charges should be linked to concrete land consumption and use. The urbanization of land brings about negative externalities and environmental damages that are going to be faced not just by those involved in the building activity but also by the next generations. Construction charges therefore should be increased according to the amount of agricultural land lost, taking into account the damage imposed on the future. One possible alternative could be to implement a local property tax (like the ICI) which varies not just according to the cadastral rent, but also with the amount of land lost. In this way the local property tax would become part of the evaluation procedure of consumers when making the decision of buying a certain apartment/ building.

5. CONCLUSIONS

Since the 1970s in the field of land management, Italy has applied a diversified package of local policies, comprehensive of market-based instruments and planning tools aimed at attaining the sustainable management of land. However, the economic crisis, the contraction of money transfers from central government and the consequent growing local budget needs have partly modified the reasoning underpinning these mechanisms. Nowadays, construction charges are mainly used to cover budget needs and, as a result, the current local fiscal regime encourages urban sprawl.

The link between urban planning and local budget revenues calls for the intervention of public authorities. A diversified package of policies should be defined so as to favour the balanced and sustainable consumption of land along with the economic sustainability of local budgets. In this field it is necessary to devise a new system capable of grasping the main environmental concerns related to land management. New market-based instruments serving environmental issues should be combined with interventions on the existing traditional fiscal instruments (like exemptions or tax rebates) at a local level with the aim of avoiding overdraft situations (Cipollina, 2008).

Several options can be identified as meeting this general purpose.

One is the repeal of the construction charge. If urbanization works were funded by the entire community, local governments would not have any incentive to enlarge the areas suitable for urbanization. Moreover, given that urbanization works increase a building's value,[18] a property tax like ICI, the local council property tax, could be accurately arranged in order to include the value of the environment surrounding the building (Osculati, Zatti, 2010).

If the option chosen is to keep the construction charges in place, they should be modeled so as to reflect the costs involved in the provision of urban public goods and prompt the efficient exploitation of land. Consumers located in the city centers should be charged less than those living at the borders, because longer distances entail higher costs of implementation for the provision of public goods and urban infrastructures. The construction charge component related to construction costs could be turned into an environmental incentive/disincentive. Specifically, it could be modified according to the nature of the materials used and projects implemented rewarding those saving energy and penalizing the others (Osculati, 2009).

At the same time, equity and distributive instruments must be implemented to lessen the problems of territorial competition. In the field of metropolitan areas, to address the fiscal competition among neighboring territories it would be better to opt for an overall metropolitan regulation and the creation of a metropolitan fund, where a certain percentage of the revenues collected by each municipality would accrue.

Finally, some suggest the adoption of a pigouvian tax, including all the relevant externalities generated by a building. However, its level would be much higher than current construction charges and would pose relevant problems of acceptability. Moreover, this tax could be afforded by just high-quality and high-value building construction, with the risk of relevant equity problems.[19]

Certainly, the realization of a proper pigouvian tax is far from being actually implemented. In the meantime, the regulatory authority should think about the modulation of the existing construction charges and the local council property tax (ICI) so as to properly account for land loss and consumption. They should be raised to represent the land loss faced not just by the current generations, but also by the future ones. The current urban charges, in fact, can be seen as short term-income, defined regardless of any mid- or long-term analysis of the economic and environmental consequences. Linking construction charges to its future effects (through an increase in the tax rate) would help countervail the problems of urban sprawl, which traditionally originate when construction charges are used to cover the current local budget needs.

NOTES

1. It is the case of lands with high hydrogeological instability.
2. In most cases the land market structure tends to be a monopoly or oligopoly since the involved goods are characterized by scarcity, uniqueness and impossible reproducibility.
3. It is the case of underproduction of infrastructures such as roads and electricity networks.
4. D.P.R. stands for 'Italian Republic Presidential Decree'.
5. Recently, some new normative regulations (D.P.R. n. 380/2001, Budget Law 2008) have lessened the role of these contributions in the funding of urbanization planning. The Budget Law 2008 established that, until 2010, 50 per cent of the revenues coming from construction charges could be used to fund current expenditures; while another 25 per cent could be used to fund the asset maintenance.
6. Notwithstanding the main aim of this component of the construction charge was to have a direct connection to the building market value, it failed to meet its objective. In fact, its level should constantly line up to the market dynamics; while it happens that when building market prices go up the level of this component is practically unchanged (Agnoletti, Ferraina, 2010).
7. These are updated every five years by each municipality in accordance with what has been established at the regional level.
8. The basic cost of urbanization for the residential sector was set at €129.11/m^2 in 1990. Today its value, applying the ISTAT revaluation coefficients, amounts to €234.88/m^2.

9. The weight of the construction charge with respect to the entire construction costs varies between 5 per cent and 20 per cent according to the building features, building uses and location (Art. 16–17 D.P.R. n. 380/2001).

10. The process of fiscal devolution launched during the Sixties and partly formalized with Law n. 662/1996 assigned jurisdictions above land planning and defense to local municipalities. According to Law n. 42/2009 Art.12 on fiscal federalism, the expenditures of local entities used to carrying out their functions can be covered by: local taxes and duties; surtaxes or possibly equalization transfers. With respect to local taxes and duties, municipalities can rely on a wide range of economic instruments: local council property tax (ICI, *Imposta Comunale sugli Immobili*); surtax on the personal income tax (IRPEF , *Imposta sul Reddito delle Persone Fisiche*); tax on the use of public soils (TOSAP, *Tassa per l'Occupazione di Spazi ed Aree Pubbliche*); tax on the collection or disposal of solid waste (Tarsu, *Tassa per lo Smaltimento dei Rifiuti Solidi Urbani*); advertisement tax, surtax on electric energy consumption, etc.

11. Budget problems have been exacerbated by the recent repeal of the local council property tax (hereinafter ICI, *Imposta Comunale sugli Immobili*) levied on the first house and the abrogation of the destination bond of the collected revenues.

12. Basically, part of the collected revenues is used to fund current expenditures, instead of capital expenditures.

13. The demand for urban infrastructures and services would probably show up after the arrival of urban development.

14. The compensation fund should be fed by transfers from the regional governments and revenues collected through construction charges.

15. However, from a legal point of view, taxing the next generations means taxing passive actors who are not responsible for what has been done in the past.

16. This situation has been exacerbated by the two amnesties for infringement of local building regulations in 1994 and 2004 and by the lack of any regular and wide urban planning.

17. The regulatory authority should first of all distinguish amongst urban, metropolitan, and rural municipalities. Secondly, population density and budget needs should also be part of the decision-making process towards the setting of new urbanization and construction charges.

18. In this way the local communities could get back part of their expenses.

19. In order to lessen these effects, the collected revenues could be used to fund public housing and other public goods (Osculati, Zatti, 2010).

REFERENCES

Agnoletti, C. and Ferraina, G. (2010), 'Il contributo di costruzione nel finanziamento della città pubblica. Il caso dell'area fiorentina', in ISAE, IRES Piemonte, IRPET, SRM, IRER (eds.), *La finanza locale in Italia. Rapporto 2009*, Milano: Franco Angeli. (ISAE – *Istituto di Studi e Analisi Economica*, Institute for Studies and Economic Analyses; IRES Piemonte – *Istituto di ricerche economiche e sociali*, Economic and Social Research Institute of Piedmont; IRPET – *Istituto regionale programmazione studi e ricerche per il Mezzogiorno*, Studies and Research Association for Southern Italy; IRER – *Istituto regionale di ricerca della Lombardia*, Lombardy Regional Institute for Research.)

Arnott, R. (1987), 'Economic theory and housing', in Mills, E.S. (ed.) *Handbook of Regional and Urban Economics*, vol. II, chapter 24, Elsevier Science Publisher.

Ave, G. (1999), 'Urbanistica e tassazione delle proprietà immobiliari in Italia', in Curti, F. (ed.), *Urbanistica e fiscalità locale. Orientamenti di riforma e buone pratiche in Italia e all'estero*, Rimini: Maggioli.

Brueckner, Jan K. (1997), 'Infrastructure financing and urban development: The economics of impact fees', *Journal of Public Economics*, **66**(3), 383–407.

Ceriani, A., Penatti, L. and Pola, G. (2008), 'Gestione del territorio, perequazione e compensazione territoriale. Ipotesi di applicazione in Lombardia', in ISAE, IRES Piemonte, IRPET, SMR, IRER (eds.), *La finanza locale in Italia. Rapporto 2007*, Milano: Franco Angeli.

Cipollina, S. (2008), 'Fiscalità e tutela del paesaggio', *Rivista di diritto finanziario e scienza delle finanze*, **1**(4), 552–72.

Curti, F. (1999), *Urbanistica e fiscalità locale. Orientamenti di riforma e buone pratiche in Italia e all'estero*, Rimini: Maggioli.

Curti, F. (2004), *Valutazione dei progetti urbani e fiscalità urbanistica*, Rimini: Maggioli.

European Environment Agency (EEA) (2006), *Urban sprawl in Europe. The ignored challenge*, EEA Report n. 10/2006.

Ferri, V. (2007), *Limitare il consumo di suolo & costruire ambiente*, Milano: Politecnico.

Gilbert, G. and Guengant, A. (1993), 'Fiscalité locale et aménagement du territoire', in Blanc J., Gilbert, G., Guengant, A. *et al.* (eds.), *Finances locales et aménagement du territoire*, Paris: Datar.

ISPRA (2008), *Qualità dell'ambiente urbano*, Rome: *Istituto Superiore per la Protezione e la Ricerca Ambientale* (Institute for Environmental Protection and Research).

Magnani, I. (2001), *Progresso tecnico, pubblico e privato, città e dintorni*, Roma: Accademia dei Lincei.

Magnani, I. and Muraro, G. (1978), *Edilizia e sviluppo urbano*, Milano: Franco Angeli.

Malpezzi, S. (1999), 'The regulation of urban development: lessons from international experience', The University of Wisconsin, *CULER Foundation Working Papers*, 99–07

Micelli, E. (2002), 'Development rights markets to manage urban plans in Italy', *Urban Studies* **39**(11), 141–54.

Osculati, F. (2009), 'Fame di terra', www.nelmerito.com.

Osculati, F. and Zatti, A. (2010), 'Costituzione e ambiente', *il Politico*, vol. 3, pp. 108–38.

PART IV

Environmental Taxation and Natural Resources

9. Coal mining: the neglected environmental threat

Hans Sprohge and Julsuchada Sirisom

Coal mining causes considerable damage to the environment. Some of the damage is irreparable. An understanding of why this is so requires understanding how coal is formed. How coal is formed explains its location in the earth's crust. Various mining methods are used to extract coal from the crust of the earth. These methods are very destructive to the environment. Both the federal government and the various states have enacted laws to protect the environment by requiring reclamation of mined areas. Sometimes these laws are ineffective in protecting the environment. They may not be enforced or skirted through legal chicanery. Even when there are good faith attempts to comply with laws intended to protect the environment, reclamation may not be possible. The damage to the environment from coal mining is exacerbated by tax incentives and externalities. Removing tax incentives for environmental degradation and creating tax incentives for environmental restoration can mitigate the negative environmental effects of coal mining. Additional benefits of doing so are that alternative energy sources may become competitive with coal and significant conservation efforts may be undertaken.

COAL FORMATION

Coal formed millions of years ago from massive accumulation of dead, land-based plant life, mainly trees.[1] The energy in coal comes from the energy that plants absorbed from the sun millions of years ago.[2] The plants that formed coal captured solar energy through photosynthesis to create the compounds that make up plant tissues. The most important element in the plant material is carbon, which gives coal most of its energy.[3] When plants die, this energy is usually released as they decay. When coal was formed, the decay process was interrupted, preventing the release of the

129

stored solar energy. The energy is locked into the coal.[4] Coal is classified as a nonrenewable energy source because it takes millions of years to form.[5]

It was not until 300 million years ago that plants developed sufficiently to form primordial forests that produced the major coal deposits.[6] Steamy swamps and bogs were everywhere. Trees and plants grew everywhere. As plants and trees died, their remains sank to the bottom of the swampy areas, accumulating layer upon layer. As each new layer of dead and dying plants increased in thickness, these vast swamps slowly sank. The first stage in the 'coalification' process is the decomposition of this debris by fungi and bacteria.[7] Proteins, starches and cellulose decompose more rapidly than the woody material and the waxy parts of the plants.[8] That is why carbonized fibers, stems, leaves, seeds of plants, and even tree stumps can be found in coal.

The rapid decaying of the plant material results in the formation of 'peat'. As the makeup of the earth's surface changed, seas and great rivers caused deposits of sand, clay and other mineral matter to bury the peat.[9] Eventually, hundreds and sometimes thousands of feet of sediment covered the peat. This burial prevented air from reaching the peat. The lack of air started the second stage of the 'coalification' process.[10] In this second stage, sandstone and other sedimentary rocks were formed. The pressure caused by the weight of the accumulated layers of overlying sediments and rock upon the submerged plant matter squeezed out much of the water and caused some of the volatile substances to escape and the nonvolatile carbon material to form a more compact mass.[11] Increasingly deeper burial and the heat associated with it gradually convert the peat first into brown coal (lignite) and then into sub-bituminous coal, bituminous coal and finally to anthracite.

COAL PRODUCTION

Coal is mined using surface and underground methods. The method used depends on depth of burial, density of the overburden and thickness of the layer of coal known as a seam. Overburden is the layers of soil and/or rock that lie over the coal. Surface mining is used for seams relatively close to the surface, at depths less than approximately 180 feet.[12] Underground mining is used for seams that occur at depths of 180–300 feet. Sometimes surface mining methods are used for seams that occur at depths greater than 180 feet. In some western states in the United States, seams with a thickness of 60–90 feet that occur at depths in excess of 200 feet are mined by open pit methods. Underground methods are usually used for seams occurring below 300 feet.

Underground mining utilizes a machine with a large rotating steel drum equipped with tungsten carbide teeth or spikes that tear coal from the seam wall.[13] In surface mining (another name for 'strip mining'), a coal seam is exposed by removing the overlying vegetation and overburden.[14] Mines in steep areas, typically have overburden made up of rock. Explosives are used to break up the overlying rock.[15] Mines in flat areas typically have overburden made up of soil, rather than rock.

A particularly pernicious form of mining is mountaintop removal. It consists of removing entire mountaintops to reach thin layers of valuable low-sulfur coal seams lying underneath the mountains of central Appalachia. Mountaintop removal generates an enormous amount of overburden. Trees are ripped from the ground and brush cleared with huge tractors.[16] This debris is burned. Deep holes are dug for explosives. Explosive are poured into these holes. As much as 800 to 1000 feet are blown off the mountain tops in order to reach a coal seams buried deep below. Unlike every other surface mining method, none of the mined area is backfilled.[17] Steep mountain grades make restoring the natural contour of the landscape impossible. Huge draglines – some the size of an entire city block, able to scoop up to 100 tons in a single load – push the overburden into nearby streams and valleys.

COAL MINING EFFECTS ON WATERWAYS

Both surface and underground mines produce acidic water, commonly known as acid mine drainage or acid rock drainage.[18] Generally, pyrite (iron sulfide), also known as 'fool's gold' and other sulfide minerals are contained in coal seams, coal storage piles, overburden, and coal processing wastes. Exposure of sulfide minerals to air and water results in the oxidation of sulfur and the production of sulfuric acid and elevated concentrations of dissolved metals, such as, iron, sulfate, and other metals. The metals stay dissolved until the pH rises to a level where precipitation occurs. Sulfuric acid production continues for as long as sulfide minerals are exposed to air and water, whether the mine is still operating or not. With surface mining, the entire exposed seam leaches sulfuric acid for as long as it rains.[19] With underground mining, sulfide materials are exposed to air and water through subsidence and rain. Mine subsidence is the movements of the ground surface as a result of readjustments of the overburden due to collapse or failure of underground mine workings.[20] Subsidence may cause surface water to be diverted underground.[21] Furthermore, with underground mining, waste materials are piled at the surface.[22] Rain percolating through these piles creates runoff that pollutes local streams. Acid mine drainage and acid rock drainage leave the subsoil infertile on the surface and pollutes streams by acidifying and killing fish, plants, and aquatic animals. Exposed acid material may continue to leach acid for 800 to 3000 years.[23]

COAL MINING EFFECTS ON PLANTS AND ANIMALS

The Ohio Valley Environmental Coalition calls mountaintop removal ecocide – and with good reason! Only the Amazon in South America rivals the Appalachians in biodiversity.[24] According to Central Appalachian edition of *The Smithsonian Guides to Natural America*, the Kanawha State Forest in West Virginia has more than 1000 species of trees and plants within its 9474 acres. Kanawha State Forest is typical of the southern West Virginia mountains. A mountaintop removal mine borders the southern edge of the forest.[25] Mountaintop removal completely destroys all plant an animal life, including several endangered species.[26] After a mountaintop is deforested, the remaining forest may become fragmented. Fragmentation breaks a forest into many pieces rather than keeping it together as a large intact tract. As a result, the exterior forest area is larger than interior forest area.[27] A decrease in the interior area of a forest limits habitat for animals, plants, and leaves it susceptible to invasive species. A decrease in habitat leaves animals more susceptible to predators or nest robbers.[28] When sections of a forest are cut and not re-planted, the species that make up that forest die alongside.[29] Fragmentation makes the forest interior more susceptible to exotic and threatening species that often destroy massive amounts of habitat when introduced into a new system; resulting in a loss of natural species biodiversity.[30]

In West Virginia, more than 300 000 acres of hardwood forests (half the size of Rhode Island) have been destroyed by mountaintop removal.[31]

From the ground, the destructiveness of mountaintop coal mining is hidden from view. A fringe of trees along roadsides hides it. However, an aerial view clearly shows the destructiveness. Go to Google Maps, enter West Virginia, and click on Satellite view.[32] Scar-like areas dotting the central southern part of the state will be visible. Center any of them in the window and enlarge it. The extent of the destruction is readily visible.

Videos of mountain top mining operations can be found using Google's advanced search function. Enter 'mountaintop mining' and in the 'Search within a site or domain:' enter '.youtube.com.' These videos show that after mountaintop removal, mountaintops become large, flat plateaus.

Mountaintop removal also destroys habitat of aquatic organisms, including many endemic species. Sometimes, the overburden that is pushed into nearby streams can completely bury and destroy the stream altogether.[33] Streams that are not completely covered can still suffer extreme degradation from mining pollution, leading to less diverse and more pollutant-tolerant species.[34]

Source: http://earthobservatory.nasa.gov/Features/MountaintopRemoval/.

Figure 9.1 Overburden pushed into a stream in West Virginia

In a 12 million-acre area in east Kentucky, south West Virginia, west Virginia, and east Tennessee, 1200 miles of streams have been damaged by mountaintop removal.[35] Valley fills can greatly increase the severity of flash flooding. By 2012, mountaintop coal mining will have serious damaged or destroyed an area larger than Delaware.[36]

LAWS TO PROTECT THE ENVIRONMENT

Various federal and state laws require coal mine reclamation. Reclamation is the rehabilitation of land after cessation of coal mining operations. The Surface Mining Control and Reclamation Act (SMCRA) enacted in 1977 was the first federal surface mining law in the United States. The SMCRA is the primary federal law that regulates the environmental effects of coal mining. It does not abolish strip mining or mountaintop removal mining. The primary goal is to 'establish a nationwide program to protect society and the environment from the adverse effects of surface coal mining operations.'[37] It sets environmental standards mines must follow while operating and achieve when reclaiming mined land. Companies must obtain permits before conducting surface mining operations. Permit applications must describe pre-mining environmental conditions and land use, proposed mining and reclamation, how SMCRA performance standards will be met, and how the land will be used after reclamation.[38] Mining companies must post a bond sufficient to cover the cost of reclaiming the site to ensure the mining site will be reclaimed even if the company fails to clean up the land for some other reason. The bond is not released until the mining site until the government has found (after five years in the East and 10 years in the West) that the reclamation was successful.[39] The SMCRA grants government regulators the authority to inspect mining operations and to enforce provisions of SMCRA or an equivalent state statute. Inspectors can issue 'notices of violation,' that require operators to correct problems within a certain amount of time; levy fines; or order that mining cease.[40]

Other federal laws that affect the coal mining industry include, but are not limited to, the following:[41]

- National Historic Preservation Act (1966): governs the preservation of historic properties throughout the United States.
- National Environmental Policy Act (1969): establishes a national policy for the environment.
- Endangered Species Act (1973): governs the protection of endangered species.
- Resource Conservation and Recovery Act (1976): governs the control of hazardous wastes.
- Clean Water Act (1977): regulates the discharge of pollutants into water.
- Clean Air Act (1990): regulates the discharge of pollutants into the air.

Additionally, states enact coal surface mining and reclamation laws and promulgate regulations that governing surface water discharge permits, construction permits, air quality permits, solid waste disposal, and mine operating permits.

LAWS DO NOT PREVENT ENVIRONMENTAL DEGRADATION

Despite laws intended to protect the environment, environmental degradation continues. It is not possible to make a thorough assessment of the reasons for continuing environmental degradation. Some of these reasons include lax enforcement, loopholes, and corporate gamesmanship. It is impossible to know the exact extent of lax law enforcement since the enactment of the SMCRA to the present day. However, it has been documented that from the time the SMCRA was enacted in 1977 to 1995, half of all federal orders for mine reclamation were simply ignored by coal companies, the government failed to assess more than $200 million in penalties, and one in four of the 25 000 coal mines established during that time were not reclaimed.[42] In Kentucky only 8 percent of postmining land use in 2008 was returned to forests – 92 percent was not.[43] Enforcement of the SMCRA by the Tennessee Division of Surface Mining was so lax that the federal government took over the administration of the state's permitting program.[44] To get around the Clean Water Act, coal companies have been using 'streamlined permits' issued by the Army Corps of Engineers in Appalachia – almost 80 percent of permit approvals since 1997.[45] Such permits are supposed to apply only to 'minor activities that are usually not controversial' and that would have only 'minimal cumulative adverse effects on the environment.' Failure to enforce federal and state mining laws and regulations continues to this day.[46] Before coal mining companies may commence operations, they must obtain permits. As part of the permitting process, coal mining companies must submit Environmental Impact Statements. Instances exist where permits have been granted when Environmental Impact Statements were inadequate or not filled out or at all.[47]

A loophole exists in the general requirement under the SMCRA that lands disturbed by mining must be reclaimed to their approximate original contour (AOC).[48] Mountaintop removal and steep slope mining operations arc exempted from the AOC requirement as long as postmining operations constitute equal or better economic or public use[s] of the land than it was before.[49] Under this exemption, prisons and a golf course have been constructed on mountaintop removal sites.[50] One such prison has been nicknamed Sink-Sink after Sing Sing, a maximum security prison of the

New York State Department of Correctional Services in the Town of Ossining. Mountains, and the ecosystems they support, have an intrinsic value that cannot be measured in monetary terms. Nevertheless, under the SMCRA economic value of the land supersedes all values. Studies by Appalachian Voices and the Natural Resources Defense Council have found that only a fraction of mountaintop removal mine sites are reclaimed for economic development projects. Of the 410 reclaimed mountaintop removal mine sites surveyed, 366 (89.3%) had no form of postmining development, excluding forestry and pasture.[51] The Lexington Herald-Leader reported that from 1999 to 2009 economic development was planned for less than 3 percent of the roughly half-million acres of land covered by surface-mining permits in Kentucky. That is less than 14 000 acres scheduled to be reclaimed for commercial, residential, industrial, or recreational development.[52]

Legal gamesmanship used to circumvent environmental laws and regulations include segmentation, contractors, and shell corporations. Segmentation is used to avoid regulatory oversight and public hearings. Compared to large mines, small mines are not subject to as much regulatory oversight and to public hearings. Segmentation is the practice of creating huge projects that are treated legally as a series of small efforts.[53] For example, the Cooper Ridge mine in Tennessee stretches for miles around the sides and tops of many mountains. There is nothing physically to indicate that legally it is multiple mines. On a permit-by-permit basis the Cooper Ridge mine consists of many mines of less than 5000 acres each. Because they are all connected to each other, they are really one massive mine. Contractors and shell corporations are used to avoid liability for reclamation.[54] In the case of contractors, a large coal-owning corporation signs a contract with a small company to mine one of the large company's coal deposits. In obtaining the permit to mine the coal, the contractor assumes the liability for reclaiming the site. For a fee, the contractor hands over the coal to the large corporation to sell. Frequently, the small contractors, chronically short on capital, cannot fulfill their SMCRA obligation to reclaim mine sites. If the small company folds, the site is left not reclaimed. In the case of shell corporations, a corporation is formed that becomes the owner of the coal.[55] The corporation applies for mining permits and takes on the responsibility for reclaiming the mine. The corporation is dissolved after the coal has been mined and sold. What is left is empty shell without assets and no one to punish for failure to reclaim the mine site. From 1980 to 1993, just two national coal companies, A.T. Massey and Island Creek, used more than 725 different contractors.[56] Most of those contractors went out of business. They left behind more than $200 million in debts in southern West Virginia and eastern Kentucky.

The SMCRA grants private citizens and grassroots groups the right to challenge government enforcement of coal-mining regulations.[57] To prevent corporate gamesmanship, environmental activists file a complaint with the Office of Surface Mining (OSM) in the U.S. Interior Department. If the OSM accepts the complaint, it 'permit blocks' the offending company. The company gets no new permits until past violations have been corrected. Permit blocking is a very effective enforcement tool. Before a company can mine anywhere, it needs a government permit. Large coal companies with millions of dollars at risk will do anything – including reclamation of abandoned mines – to get back in business.

RECLAMATION DOES NOT RESTORE THE ENVIRONMENT

Because of lax law enforcement and unsuccessful efforts to enforce the law, the rate of coal mine reclamation occurring today is shockingly low. Reclaiming areas that have been subjected to mountaintop removal mining is not even in the realm of possibility. The destruction of all forms of habitat is complete and permanent. When reclamation does occur, it is frequently inadequate and fails to mitigate environmental damage as required by the SMCRA. The SMCRA requires mine reclamation to establish:

> ... diverse, effective, and permanent vegetative cover of the same seasonal variety native to the area of land to be affected and capable of self-regeneration and plant succession at least equal in extent of cover to the natural vegetation of the area; except, that introduced species may be used in the revegetation process where desirable and necessary to achieve the approved postmining land use plan.[58]

Reclamation of many mining sites re-vegetate using native plants but not of the 'same seasonal variety.' Frequently, though, mining sites are re-vegetated with plants that are not native to the area. Sometimes mining sites are re-vegetated with invasive species.[59]

In the western United States, non-native species are used for mine reclamation. Vegetation indigenous to the western United States has adapted to the arid climate.[60] Diverse native plant species provide forage for animals. In many areas, the annual average rainfall is less than ten inches. This lack of moisture makes re-vegetation of mined lands with native species difficult and expensive. Consequently, many mining companies use non-native species that often are not suitable for forage and may not be capable of long-term self-regeneration. Frequently, adapted agronomic plants used to reclaim mines in mountainous areas are invasive, restrict natural succession, and replace native species in previously undisturbed plant communities. In the middle-west and eastern United States grasses and legumes frequently are used in lieu of reforestation. Intensive earth moving prior to re-vegetation compacts soil.[61] Compacted soil limits root penetration and prevent proper movement of air and water into the soil. Shallow-rooted grasses can grow under these conditions, but the conditions will not benefit deep-rooted plants such as trees. Competition for soil moisture is keen in such shallow root zones. Grasses will always out-compete trees for available moisture. Grasslands resulting from reclamation provide habitats for rabbits, voles, and woodchucks. Deer from neighboring woodlands are attracted to the agronomic grasses and legumes. Deer, voles, and rabbits severely damage trees planted in these grasslands. Some tree plantings fail completely; the survival of others is reduced below acceptable limits.

Even if a coal mining company intends to reclaim waterways damaged by acid mine drainage, it may not be able to do so. A survey of 12 reclaimed sites in the Sewanee coal seam of south-eastern Tennessee found continued acid mine drainage at ten of them. Acid mine drainage can continue indefinitely, causing environmental damage long after the mine operation has ended. Acid mine drainage still occurs from hard-rock mines in Europe that were worked by ancient Romans prior to A.D. 476. Many mines may require water treatment for decades or even hundreds of years in perpetuity.

TAX INCENTIVES FOR ENVIRONMENTAL DEGRADATION

The mining industry has been, and continues to be, provided with economic incentives to wreak havoc on the environment. Chief among them are tax breaks. A study by the Environmental Law Institute shows that from 2002 to 2008 coal mining companies received billions of dollars in tax breaks annually.[62] Among the many tax breaks are the following:

- tax credits for the production of coal coal-based synthetic fuels[63]
- characterizing coal royalty payments as capital gains[64]
- exclusion from fuel excise tax[65]
- expensing (rather than capitalizing and amortizing) of the costs of surface stripping, and construction of shafts and tunnels[66]
- deductibility 10 percent of gross income from coal production[67]
- deductibility of reclamation and closing costs immediately when

beginning mining rather than when the mine site is closed, restored and the costs associated with these activities paid.[68]

The coal mining industry also benefits indirectly through the Low Income Home Energy Assistance Program. This provides low-income households with the means to make their utility payments, a good portion of which is energy generated by from coal. The Black Lung Disability Trust Fund (BLDTF) pays health benefits to coal miners afflicted with 'black lung' disease. The BLDTF is funded through an excise tax on coal. However the tax is not sufficient to cover all costs. The BLDTF was given 'indefinite authority to borrow' from the U.S. General Fund. By the end of the 2008 financial year, the BLDTF was nearly $13 billion in debt. In 2008, Congress partially 'bailed out' the BLDTF. The Environmental Law Institute tabulated this bailout as a subsidy to coal companies.

RECOMMENDATIONS

The Internal Revenue Code can be used as an instrument of environmental policy to mitigate the environmental destruction caused by coal mining operations. On the one hand, taxpayer subsidies of coal mining operations can be eliminated. On the other hand, the cost of reclamation can be shifted from the taxpayer to the mining companies. The tax preferences granted to mining companies are in effect a taxpayer subsidy. The federal and state governments eliminate direct and indirect taxpayer subsidies of coal mining companies by (a) not adding more tax preferences that lower the after tax cost of producing coal and (b) repealing those that are currently in effect. Examples of direct taxpayer subsidies are characterizing ordinary income as capital gains and expensing, rather than capitalizing, and so on. Examples of indirect taxpayer subsidies are tax credits for clean coal technology and converting coal to liquid fuels. Promoting more coal use without also providing additional environmental safeguards will only increase the toxic abuse of ecosystems.

The cost of reclamation can be shifted from the public to the mining companies by imposing taxes on carbon emissions, the value of lumber destroyed, lost carbon sinks, and lost plant and animal species. The carbon emitted into the atmosphere by coal mining operations is measurable. A tax should be imposed on these carbon emissions. The mining companies should not be able avoid this carbon tax. Such a tax would provide incentives for coal mining companies to reduce their carbon footprint. In surface mining operations, timber is destroyed. A tax should be imposed on the fair market value of the timber destroyed. The destruction of forests results in lost carbon sinks. The amount of carbon sequestration by forests is measurable. A tax should be imposed on the lost carbon sink. A lost carbon sink tax should remain in effect until reclamation has re-established carbon sequestration to pre-mining levels. If reforestation takes 20 years, then the tax should be imposed for 20 years. A lost sequestration tax would discourage the use of grasses instead of trees. Why? Because trees store more carbon than grass stores.[69] Similarly, a tax should be imposed on each type of plant and animal species destroyed by coal mining until it is re-established through reclamation. A lost species tax would discourage the use of invasive and non-native species to reclaim forests. Invasive species prevent native species from re-establishing themselves. Using non-native species would not eliminate the tax.

From a tax standpoint mountaintop removal mining can be rendered unfeasible replacing the current highest best economic use reclamation standard with a like kind and quality standard. The economic standard allows mining companies to push externalities off onto the public. A like kind and quality standard would require rebuilding mountains that were blown apart, clearing up waterways that were destroyed, and re-establishing plant and animal habitat reasonably close to its former state. Imposing taxes on lost mountaintops, streams, and habitat until they are re-established makes mountaintop removal mining economically unfeasible.

Implementation of any of the above recommendations would result in a higher coal prices. Higher coal prices would benefit alternative energy sources and lead to conservation. Coal is cheap because coal mining companies are subsidized by taxpayers and they do not bear the full cost of remediating environmental damage caused by their operations. Unsubsidized coal prices that include the cost of environmental remediation would make clean renewable technologies like wind, geothermal power, and solid biomass cost competitive with coal. Higher coal prices also would encourage conservation. Does the country really need all those neon lights? Maybe households would turn off the television when it is not being watched or the computer when it is not being used, etc. With currently available energy efficiency and conservation measures, the country could save from 20 to 30 percent of its energy usage.[70]

CONCLUSION

Coal formation dictates whether surface and underground mining methods are used. Both types of mining methods are harmful to the environment. The myriad of laws intended to protect the environment are ineffective in doing so because of lax enforcement or legal gamesmanship. Sometimes environmental degradation from coal mining is so acute that reclamation is not even possible. Coal production is subsidized by taxpayers and coal mining companies can shift the economic burden of reclaiming the environment to the public. The negative environmental impact of coal mining can be mitigated by eliminating taxpayer subsidies of coal mining and by providing an incentive to restore the environment by taxing coal mining companies until they do. The resulting increase in coal prices may have two additional environmental benefits: Alternative fuels may become competitive with coal and significant conservation efforts may be undertaken. As it is, coal is cheap, but the cost is high.

NOTES

1. planete-energies.com 'Coal Formation', www.planete-energies.com/content/coal/formation.html, accessed 6 September 2010.
2. National Energy Education Development Project 'Energy Infobooks: Coal', www.need.org/needpdf/infobook_activities/SecInfo/CoalS.pdf, accessed 7 September 2010.
3. Kentucky Educational Television 'Electronic Field Trip to a Coal Mine', www.ket.org/Trips/Coal/AGSMM/AGSMMhow.html, accessed 7 September 2010.
4. National Energy Education Development Project, *supra*, n. 2.
5. *Ibid.*
6. Powerworks 'The Formation of Coal', www.powerworks.com.au/chemistry.pdf, accessed 8 September 2010.
7. *Ibid.*
8. *Ibid.*
9. Kentucky Educational Television, *supra,* n.3.
10. Powerworks, *supra*, n. 6.
11. Christian Ngô and Joseph B. Natowitz, Our Energy Future: Resources, Alternatives, and the Environment 57 (2009).
12. Wikipedia 'Coal Mining', http://en.wikipedia.org/wiki/Coal_mining, accessed 1 September 2010.
13. *Ibid.*
14. Mark Squillace 'The Environmental Effects of Strip Mining', http://sites.google.com/site/stripmininghandbook/chapter-2–1, accessed 10 September 2010.
15. American Coal Council 'Coal Basics', www.clean-coal.info/drupal/pubs/coal_basics.pdf, accessed 9 September 2010.
16. Earthjustice 'What is Mountaintop Removal?', www.earthjustice.org/features/campaigns/what-is-mountaintop-removal-mining, accessed 22 August 2010.
17. Squillace, *supra*, n. 14.
18. Bruce G. Miller, Coal Energy Systems 82 (2005).
19. Wikipedia 'Environmental effects of coal', http://en.wikipedia.org/wiki/Environmental_effects_of_coal, accessed 1 September 2010.
20. Pennsylvania Department of Environmental Protection 'What Is Mine Subsidence?', www.dep.state.pa.us/MSI/WhatIsMS.html, accessed 17 September 2010.
21. Miller, *supra*, n. 18 at 84.
22. Martha Keating, Cradle to Grave: The Environmental Impacts from Coal 2 (2001). Available at: www.catf.us/resources/publications/files/Cradle_to_Grave.pdf.
23. Squillace, *supra*, n. 14.

24. Heidi Stevenson 'Mountaintop Coal Mining – The Greatest Environmental Sin of All', www.gaia-health.com/articles/000034-Mountaintop-Coal-Mining.shtml, accessed 18 September 2010.

25. Ohio Valley Environmental Coalition 'OVEC's "Reclamation" Galleries', www.ohvec.org/galleries/reclamation/index.html, accessed 25 September 2010.

26. M. A. Palmer, E.S. Bernhardt, W.J. Schlesinger, *et al.*, *Mountaintop Mining Consequences* 327 Science 148 (2010).

27. Sustainable Forest Partnership 'How is forest fragmentation affecting you?', http://sfp.cas.psu.edu/fragmentation/how.htm, accessed 4 September 2010.

28. Cornell Lab of Ornithology 'What is Forest Fragmentation and Why is it Important?', www.birds.cornell.edu/bfl/gen_instructions/fragmentation.html, accessed 18 September 2010.

29. Sustainable Forest Partnership, *supra*, n. 27.

30. *Ibid.*

31. Union of Concerned Scientists 'Environmental impacts of coal power: fuel supply', www.ucsusa.org/clean_energy/coalvswind/c02a.html, accessed 5 August 2010.

32. Stevenson, *supra*, n. 24.

33. U. S. Environmental Protection Agency 'Mid-Atlantic Mountaintop Mining', www.epa.gov/Region3/mtntop/#impacts, accessed 22 July 2010.

34. *Ibid.*

35. NASA 'Coal Controversy in Appalachia', http://earthobservatory.nasa.gov/Features/MountaintopRemoval, accessed 22 July 2010.

36. James Hansen 'A Plea to President Obama: End Mountaintop Coal Mining', http://e360.yale.edu/content/feature.msp?id=2168, accessed 4 August 2010.

37. Aurora Lights 'What is Mountaintop Removal?', http://auroralights.org/map_project/theme.php?theme=mtr&article=20, accessed 20 September 2010.

38. Wikipedia 'Surface Mining Control and Reclamation Act of 1977', http://en.wikipedia.org/wiki/Surface_Mining_Control_and_Reclamation_Act_of_1977, accessed 20 September 2010.

39. *Ibid.*

40. *Ibid.*

41. American Coal Foundation 'Coal Mining in America: Federal and State Regulations', www.teachcoal.org/aboutcoal/articles/coalamer.html, accessed 23 September 2010.

42. The CBS Interactive Business Network 'Cracking down on mining pollution – environmental lawyer Thomas Galloway develops Applicant/Violator System to find violators of mining law', http://findarticles.com/p/articles/mi_m1169/is_n4_v33/ai_16971135, accessed 22 September 2010.

43. Rob Perks 'Appalachian Heartbreak', www.nrdc.org/land/appalachian/files/appalachian.pdf, accessed 23 September 2010.

44. John Nolt 'Strip Mines', http://web.utk.edu/~nolt/radio/stripmn.htm, accessed 30 September 2010.

45. World Socialist Website 'The social crisis in Appalachia', www.wsws.org/articles/2010/jul2010/app3-j27.shtml, accessed 1 October 2010.

46. Upper Michigans Source.com, 'Obama nominee sited in Intent to Sue', www.uppermichiganssource.com/community/press_release.aspx?id=355771, accessed 30 September 2010.

47. Aurora Lights, *supra*, n. 37.

48. 30 U.S.C. § 1265(b)(3).

49. 30 U.S.C. § 1265(c)(3)(A), 1265(e)(3).

50. Aurora Lights, *supra*, n. 37.

51. SourceWatch 'Coal mine reclamation', www.sourcewatch.org/index.php?title=Coal_mine_reclamation, accessed 20 September 2010.

52. Bill Estep and Linda J. Johnson 'Mountains of potential?', www.kentucky.com/2009/10/18/981954/mountains-of-potential.html, accessed 25 September 2010.

53. Kari Lydersen 'Resisting Mountaintop Removal in Tennessee', www.alternet.org/environment/28489?page=4, accessed 24 September 2010.
54. The CBS Interactive Business Network, *supra*, n. 42.
55. *Ibid.*
56. *Ibid.*
57. SourceWatch, *supra*, n. 51.
58. 30 U.S.C. § 1265(b)(19) (2008).
59. United States Environmental Protection Agency 'Mountaintop Mining/Valley Fill Environmental Impact Statement', http://wvgazette.com/static/series/mining/reports/EIS/Executive%20Summary.pdf, accessed 27 September 2010.
60. Squillace, *supra*, n. 14.
61. Society of American Foresters 'Tree Planting On Strip – Mined Land', www.ohiosaf.org/planting.htm, accessed 26 September 2010.
62. Environmental Law Institute 'Estimating U.S. Government Subsidies to Energy Sources: 2002–2008', www.elistore.org/Data/products/d19_07.pdf, accessed 1 October 2010.
63. IRC Section 45K.
64. IRC Section 631(c).
65. IRC Section 6426(d).
66. IRC Section 617.
67. IRC Section 613.
68. IRC Section 486.
69. Hans Joosten 'Peat and climate: on words, facts and choices', www.imcg.net/docum/peatandclimate.ppt, accessed 3 October 2010.
70. Ohio Valley Environmental Coalition, *supra*, n. 25.

10. Great Lakes water quality and restoration programs

Rahmat O. Tavallali and Paul J. Lee

INTRODUCTION

The Great Lakes, consisting of Erie, Huron, Michigan, Ontario and Superior, are the world's largest surface freshwater source. The combined surface area of the lakes is approximately 94 250 square miles(en.wikipedia.org) over eight states, Illinois, Indiana, Michigan, Minnesota, New York, Ohio, Pennsylvania, Wisconsin, and two Canadian Providences of Quebec and Ontario. Four of the five lakes form part of the Canada-United States border; the fifth, Lake Michigan, is contained entirely within the United States.

Water levels on all five Great Lakes trended downward in the last decade to near-record lows before rebounding some in recent years. That has been attributed to warming trends, and warmer lakes could worsen that problem through greater evaporation (Condon, 2010). Humans have also contributed to the quality of the Great Lakes. Ways that humans have affected the quality of the Great Lakes water over the centuries include sewage disposal, toxic contamination through heavy metals and pesticides, overdevelopment of the water's edge, runoff from agriculture and urbanization and air pollution (Great-Lakes.net). Other sources of pollution include degradation of phosphorus, oil and hazardous polluting substance from vessels wastes and pollution from shipping sources and activities.

The United States Congress first addressed water pollution issues in the Rivers and Harbors Act of 1899. Portions of this law remain in effect, including the Refuse Act, while others have been superseded by various amendments, including the 1972 Clean Water Act.

In order to restore and protect the Great Lakes water quality, in 1972, Richard Nixon signed the first US clean water law. As amended in 1977, 1981 and 1987, this law became commonly known as the Clean Water Act. This act was a key step in reducing Great Lakes pollution and aimed to

attain a level of water quality that 'provides for the protection and propaga-
tion of fish, shellfish and wildlife, and provides for recreation in and on the
water' by 1983 and to eliminate the discharge of pollutants into navigable
waters by 1985 (Kenney, 2008). Since then a number of programs by
federal, states and local government agencies have been created to care for
the Great Lakes.

In 2005 the Great Lakes Regional Collaboration Strategy was developed
between United States and Canada. As a result, the water quality of the
Great Lakes improved significantly.

In this chapter the authors review different federal and state tax and
financing incentives programs to reduce and to end water pollution and
sewer outflows throughout the Great Lakes region. The paper also examines the effectiveness of these programs in improving Great Lakes water
quality.

COLLABORATION STRATEGY

Although releases of toxic pollutants have been reduced significantly over
the years, there is a legacy of contamination in sediments and fish throughout the system, and mercury and other pollutants continue to enter the
Great Lakes from nearby and distant sources.

> A key piece of the puzzle was put into place when President Bush issued an
> Executive Order in May 2004. This Order recognized the Great Lakes as a
> 'national treasure' and created a Federal Great Lakes Interagency Task Force to
> improve federal coordination on the Great Lakes. The Order also directed the
> Environmental Protection Agency to convene a 'regional collaboration of
> national significance for the Great Lakes.' This collaboration process was needed
> to develop, by consensus, the national restoration and protection action plan for
> the Great Lakes. (Great Lakes Regional and Collaboration, 2005)

THE COLLABORATIVE PROCESS

In December 2004, the region's leaders kicked off the Great Lakes Regional
Collaboration (GLRC). The objective of this GLRC is to 'Restore and
protect the Great Lakes.' The representatives on the Executive Committee
are federal government cabinet officials, Great Lakes governors, mayors
and tribal leaders. Representatives of Congress of the Canadian government also serve as observers. Since then, the Collaboration has developed a
strategy that provides a set of recommendations to restore and protect this
national treasure.

The 2004 GLRC strategy was based on each team's highest priority recommendations for actions that can be taken over a period of time to effectuate improvements in the Great Lakes Basin (glrc.us/document). A series of recommendations for actions and activities aimed at starting the restoration of the Great Lakes ecosystem over the next five years was drawn up.

FUNDING

The total cost of implementing the GLRC strategy was estimated to be $20 billion over five years. The breadth and estimated cost of the strategy have led some to contend that implementing all the programs outlined in the strategy would not be financially feasible within the five-year time span the strategy covers. Some suggest that setting priorities for implementing programs within each element would help decision makers choose which programs to fund, if funds are limited. Some priorities discussed at a congressional hearing on the Great Lakes Strategy included restoring and protecting near-shore and coastal waters of the Great Lakes, controlling aquatic invasive species and addressing the problems of nonpoint source pollution (ncseonline.org). However, critics point out that current funding levels and federal manpower commitments are not sufficient to address the increased needs for environmental and social data to support the Great Lakes ecological protection and restoration needs. They believe that without additional investment to support more cohesive and comprehensive implementation, the current observation/monitoring status may actually decline, since operational costs constantly increase (glrc.us/document).

Although a suite of Great Lakes indicators has been developed through State of the Lakes Ecosystem Conference (SOLEC) process, few monitoring programs are currently funded to provide necessary data for the indicators to be fully utilized.

OTHER FUNDING AND TAX INCENTIVES

In 1989 the Governors of the Great Lakes states created the Protection Fund to help them protect and restore their shared natural resources. It is a source of financial support for groups that value innovation and entrepreneurship focusing on tangible benefits for the Great Lakes ecosystem. The fund is exempt from income taxes under Section 115(1) of the Internal Revenue Code and applicable state law.

From its inception through December 2007, the fund has made a total of 216 grants and program-related investments, representing a $53 million commitment to protecting and restoring the ecological health of the Great Lakes ecosystem. (Great Lakes Protection Fund Annual Report, 2007).

President Obama, as a part of his 2010 fiscal year budget, proposed $475 million in funding to fully implement the collaboration strategy and to address issues that affect the Great Lakes, such as invasive species, non-point source pollution and toxic and contaminated sediment.

Funds are expected to be allocated strategically to implement both federal programs and projects implemented by states, tribes, municipalities, universities, and other organizations are specifically targeted at the following priorities (Great Lakes Commission, 2009):

- cleaning up toxic substances and Areas of Concern ($147 million)
- Preventing or removing aquatic invasive species ($60 million)
- improving nearshore health and preventing nonpoint source pollution ($98 million)
- restoring and protecting habitat and wildlife ($105 million)
- evaluating and monitoring progress ($65 million).

THE EFFECTIVENESS OF THE GREAT LAKES RESTORATION PROGRAMS

According to his testimony before the subcommittee on Water Resources and Environment, Committee on Transportation and Infrastructure, House of Representatives, John B. Stephenson, Director of Natural Resources and Environment, there are 33 federal and 17 state-specific environmental restoration funding programs in the Great Lakes Basin employing a variety of activities to address specific environmental problems, but there is no overreaching plan for coordinating these disparate strategies and program activities into a coherent approach for attaining overall basin restoration goals (Stephenson, 2004). The key issue is that each state measures and monitors different standards and then reports those in a different way around the basin, making determining the health of the Great Lakes more difficult (Kumble and Brabec, 2005).

Additionally, the lack of consistent, reliable information and measurement indicators makes it impossible to comprehensively assess restoration progress in the Great Lakes Basin (Kumble and Brabec, 2005).

The need for better coordination, integration and enhancement of observing and monitoring activities has been recognized for quite some time. To achieve this goal and in order to protect and restore the Great

Lakes water quality, it requires a comprehensive and consistent strategy and detailed quantitative data analysis by integrating information network and with sufficient fund and support by each of eight states.

CONCLUSION

The Great Lakes – Superior, Michigan, Huron, Erie and Ontario – is the largest group of freshwater lakes on earth and provide drinking water for more than 40 million people. The Great Lakes are critical to the health and welfare of the waterway, fish, wildlife as well as cargo traffic and recreation in and on the water of both the United States and Canada. Since 1972, the inception of the first United States Clean Water Act, a variety of steps and activities as well as grants and tax incentives programs by local, states and federal governments have been taken to reduce industrial and municipal pollution discharges into the Great Lakes. As a result the quality of water had been improved significantly. However, there is a growing concern that urban runoff and sprawl, raw sewage disposal, agricultural and toxic industrial will continue to threaten the environmental health and overall quality of life in the Great Lakes Basin.

In addition, the current funding levels and lack of a comprehensive strategy, operational information systems, and coordination between agencies involved along with inadequate monitoring activities have resulted in the full scale of water quality not been fully achieved. Each state has funded heavily on its own specific needs. There is no doubt that a regional coordination body is needed to ensure communication, cooperation and planning for all these activities within the region. Without such a coordination and interagency communications and information exchange, it is difficult to determine the effectiveness of these many environmental programs and organizations.

REFERENCES

Clean Water Primer #1, Law, Lake Ontario, www.waterkeeper.ca/documents/pri, accessed 21 July, 2010.
Condon, P. (31 July 2010), 'Warming of Great Lakes Worrisome' The Associate Press, Canton Repository, A-8.
Great Lakes Commission, May 2009, cited at www.michigan.gov/documents/deg.
Great Lakes Regional and Collaboration (2005) 'A strategy to restore and protect the Great Lakes', www.sehn.org/pdf/great_lakes.pdf., accessed 7 August, 2010.
'Great Lakes Protection Fund, 2007 Annual Report', www.glpf.org/about/07annual report.pdf. accessed 10 August, 2010. http://en.wikipedia.org/wiki/Great_Lakes.

Kenney, R. (2008), 'Clean Water Act, United States', www.eoearth.org/article/Clean_Act_United _States, accessed 9 August, 2010.

Kumble, P. and Brabec, E. (2005) 'Land Planning and Development Mitigation for Protecting Water Quality in the Great Lakes System: An Evaluation of U.S. Approaches'. Urban Rural Interface Conference Proceedings, http://works.bepress.com/elizabeth_brabec, accessed 10 July, 2010.

'Regional Collaboration Indicators and Information Strategy Team Report', Page 2, www.glrc.us/documents/strategy/I&I-appendix.pdf. accessed 3 August, 2010.

Stephenson, J. (2004) 'A comprehensive Strategy and Monitoring System Are needed to Achieve Restoration Goals' United States General Accounting Office, GAO-04–782T.

'Water Pollution in the Great Lakes', 'www.great-lakes.net/teach/pollution/water/water1.html, accessed 3 August, 2010.

www.glrc.us/documents/strategy/I&I-apendix.pdf., accessed 7 July, 2010.

www.ncseonline.org/NLE/CRSreports/08Feb/RL33411.pdf. accessed 7 August, 2010.

11. The use of market based mechanisms to bolster forest carbon

Celeste M. Black[1]

I. INTRODUCTION

Forestry activities have an important role to play in addressing the challenges of climate change. Deforestation contributes significantly to greenhouse gas emissions[2] whilst afforestation and reforestation provide an opportunity to remove carbon from the atmosphere through the establishment of carbon sinks.[3] Market-based mechanisms can be used in a variety of ways to encourage the establishment of new forests. Afforestation projects can generate carbon offsets under the Clean Development Mechanism (CDM) and domestic-level incentives can include generous tax treatment of expenses, direct grants and the generation of forest offsets. The conservation of existing forests can also be supported. One example of a measure with significant promise is the proposed Reducing Emissions from Deforestation and Degradation (REDD) mechanism.[4] This chapter focuses on market mechanisms that have been developed at a domestic level in Australia and New Zealand as these jurisdictions have developed a range of different approaches to incentivize the establishment and maintenance of forests.[5]

By way of background, the chapter commences with an overview of the potential role of forestry in responding to climate change. It then analyses the market-based incentives developed in Australia and New Zealand to encourage forestry establishment and maintenance. A critical comparison of the mechanisms follows.

II. BACKGROUND TO THE ROLE OF FORESTRY

A.   Forests as Carbon Sinks

Forests are recognised as 'one of the most important storehouses of carbon.'[6] It is estimated that existing forests store more carbon than is in the atmosphere.[7] Of the exchange of carbon between the atmosphere and the land each year, 90 per cent is attributed to forests,[8] with other natural mechanisms being the sequestering of CO_2 in soil[9] and ocean uptake of carbon.[10] Although trees release carbon during their lifetime, because the rate of carbon absorption and storage by forests is higher than release over the life of the forest, trees are recognised as a net sink of carbon.[11]

The rate at which a forest absorbs and stores carbon depends on a combination of factors, including tree species, characteristics and growing conditions such as temperature and forest management.[12] It is generally recognised that there is a positive correlation between a tree's growth and its carbon capture potential.[13] It also appears to be accepted that trees reach a carbon saturation point and stop capturing carbon at maturity.[14] Due to the relatively lengthy period over which trees generally mature (some taking up to 100 years), carbon sequestration through reforestation should be seen as a medium-term contribution to achieving lower atmospheric carbon levels and should ideally be used in combination with measures to encourage the maintenance of existing forests.

It has been argued that forests continue to be valuable after maturity due to their ability to provide long-term carbon storage,[15] but stored carbon is largely released upon harvest[16] and for this reason plantation forestry is not considered to be a form of carbon sequestration. However, many 'permanent forest' mechanisms do allow for small-scale harvesting on the assumption that new growth in the forest will maintain the level of sequestration.

B.   Global Markets

Forestry's role in addressing climate change formed part of the original Kyoto Protocol negotiations and both the UN Framework Convention on Climate Change and the Kyoto Protocol contain recognition of this role. However, the negotiations were plagued by controversy and scientific uncertainty.[17] The Kyoto Protocol provides that Annex I parties 'may rely on domestic reductions in greenhouse gas emissions resulting from forestry activities, limited to afforestation and reforestation since 1990' and upon afforestation and reforestation projects undertaken in developing countries under the CDM to meet their emissions reduction targets.[18] However, despite this recognition, international efforts to address climate change have primarily focused on the optimal domestic measures for achieving emissions reductions within particular sectors, mainly the industrial and energy sectors.[19] The lesser emphasis on forestry is due to a combination of factors, including the view that forestry measures play only a supporting role to primary mechanisms, concerns around parity between developed and developing countries (specifically whether developed countries should be able to offset emissions in particular sectors by undertaking forestry activities), concerns around monitoring and measurement and the issue of permanence.[20] In more recent years, as techniques for monitoring and measurement have developed, there is growing recognition of the potentially significant contribution of forest carbon sinks to an overall strategy to manage atmospheric greenhouse gas concentrations[21] and this is reflected in the increased number of forest incentives enacted and proposed in various jurisdictions.

The global offset market also shows the increasing value placed on forestry activities. A recent survey of the forest carbon market notes a number of significant factors in the upward trend in the volume of forest carbon offsets traded, including the fact that nine afforestation/reforestation CDM projects received registration in 2009 while only one similar project had been previously registered.[22] Specific CDM project guidelines have now been developed for such projects[23] and seven were registered in 2010.[24] Another significant factor affecting the market has been the entry of forestry backed Assigned Amount Units (AAUs) generated by New Zealand forestry activities.[25] Purchasers of voluntary offsets see forest offsets as highly desirable, especially when those projects are located in developing countries.[26]

C.   Market Mechanisms for Forestry

Market-based incentives can be key policy instruments when used by government to encourage particular behaviour. For incentives to discourage deforestation or encourage reforestation to have an impact, the incentive must provide a benefit which increases the comparative advantage of conserving or establishing a forest relative to using the land in other ways.[27] Alternative land uses may include harvesting or plantation forestry or converting the land to agricultural use whereas maintaining a forest may not directly produce an economic return. A market-based mechanism can provide the price signal needed to support the forestry activities and thereby reflect the value of public good of carbon sequestration.[28]

Incentives are commonly viewed as involving two broad types. Direct incentives are designed to have a direct impact on investment and can take the form of grants and concessional tax treatment.[29] Indirect incentives involve changing factors that are external to the particular investment and the forestry sector but which make investment in forestry more attractive.[30] A key mechanism in the forestry sector has been the issuance of carbon offsets that can be sold into an offset market. Examples of many of these forms of incentives can be found by analysing the forestry incentives available in Australia and New Zealand.

III. FOREST INCENTIVES IN AUSTRALIA

In Australia there are currently several regimes spanning the federal and state levels that aim to encourage the establishment of new forests. The main incentive at a federal level comes in the form of accelerated deductibility of forest establishment costs and a component of the proposed Carbon Pollution Reduction Scheme (CPRS) would allow for domestically generated forestry offset credits to be used to meet compliance obligations under the scheme. The federal government also recently produced standards for voluntary carbon neutral program which acknowledge the role of forest offsets.[31] At the state level, New South Wales operates an emissions trading scheme that provides a mechanism for the generation of forestry-based abatement certificates.[32] These various mechanisms are described in detail below.

A. Federal Level: Carbon Sink Establishment Deductions

The federal government has only recently, and in a relatively minor way, provided incentives for carbon sink forests. In contrast, incentives have existed for some time to encourage plantation forestry as part of a broader policy agenda.[33] Investment in plantation forestry has been encouraged by the use of managed investment schemes (MISs), which allow for the pooling of investment funds in order to finance the carrying on of the activity, and through the more generous tax treatment of expenditure under these schemes.[34]

In 2008, the federal government introduced a new regime into the federal income tax to specifically allow deductions for capital expenditure incurred in establishing a carbon sink forest where these expenses would not otherwise be deductible.[35] For the 2007–12 period, capital expenditure incurred in establishing trees in a carbon sink forest is immediately deductible and qualifying expenditure incurred in later income tax periods is depreciated more gradually at the rate of 7 per cent per annum.

A number of requirements must be satisfied to qualify for the deduction. One requirement is that the taxpayer must be carrying on a business and the trees must be established for the principal and primary purpose of carbon sequestration.[36] Expenditure must be 'establishment expenditure', such as the cost of acquiring and planting seeds or plants,[37] but does not include the cost of acquiring the land.[38] The trees must occupy a continuous land area of 0.2 hectares or more and must be likely to attain a crown cover of 20 per cent or more and reach a height of at least 2 metres, which is consistent with the definition of a forest for the purposes of reforestation under international accounting frameworks.[39]

The taxpayer must advise the Australian Taxation Office (ATO) of certain details in relation to the carbon sink forest,[40] including a brief statement of how the taxpayer expects the trees to meet conditions relating to density, canopy cover and compliance with environmental and natural resource guidelines relating to managing trees for carbon storage (found in Environmental and Natural Resource Guidelines).[41]

A deduction may be denied if the Climate Change Secretary notifies the ATO that the relevant carbon sink forest has not complied with the conditions relating to density, canopy cover or the environmental and natural resource guidelines relating to managing trees for carbon storage but the mechanism to effect this denial is unclear.[42] The legislation is also unclear on the issue of monitoring and no extrinsic material issued to date provides this information.

It is unlikely that providing favourable tax treatment to establishment expenditure will alone provide sufficient incentive for reforestation. Another mechanism is necessary whereby investors could obtain a return on their investment.[43] Currently, certain forestry activities can generate offsets through the New South Wales (NSW) GGAS scheme (described below), but this is not available on a national level. The proposed CPRS would have provided such a mechanism and it is suggested that the carbon sink deductions were introduced in anticipation of this.

B.   Proposals Under the CPRS

The CPRS is a package of proposed (but deferred) measures designed to assist Australia to adjust to a climate constrained economy where a key component is a broad-based cap and trade scheme. The proposed regime would allow entities, on a voluntary basis, to receive free Australian emissions units for undertaking qualifying reforestation activities.[44] A policy decision was made not to include deforestation in the CPRS.[45] Much of the detail of the regime for reforestation was to be provided in regulations but, as the introduction of the CPRS was deferred until after the current commitment period of the Kyoto Protocol at the earliest,[46] these regulations were never released. However, a discussion paper provides some insights into potential approaches under the CPRS regulations.[47]

Broadly, to qualify as reforestation under the regime, an entity would need to establish, manage and maintain one or more forest stands in an area declared to be a project area.[48] Forest stand is defined consistently with Australia's definition of forest in the Kyoto Protocol and the carbon sink deduction.[49] It was assumed that the requisite crown coverage should be reached within 10 years of establishment.[50]

Although the federal government considered a system whereby units would be issued or required to be surrendered based on the net carbon emissions and reductions for each year, it decided to apply an averaging approach to unit crediting based on a five-year reporting cycle.[51] A limited number of units that could be issued per project (the reforestation unit limit) would be determined by the Australian Climate Change Regulatory Authority (the Authority) upfront, based on an assessment of the project's potential for removing greenhouse gas emissions.[52] It was contemplated that harvesting would be allowed and the unit limit for harvested forest stands would be based on an average of cumulative net removals, taking into account positive removals from sequestration and negative removals from harvest, over the long term, where this would likely reflect the commitment period of 100 years.[53] The unit limit for non-harvest forest stands would be based on the total projected removals less a buffer to manage the risk of reversal of sequestration due to natural events (the risk of reversal buffer).[54]

The Authority would have the power to revoke an established reforestation project if any of the conditions required to qualify as a project are not met.[55] If a reforestation project were revoked, the Authority could require the entity to relinquish units issued in relation to the reforestation project.[56]

C.   Forestry Incentives at the State Level: NSW GGAS

In 2003, the NSW state government introduced a state wide mandatory emissions trading scheme for the electricity sector called the Greenhouse Gas Reduction Scheme (GGAS).[57] The GGAS is a baseline and credit scheme whereby an annual state-wide 'CO_2-equivalent' (CO_2-e) emissions benchmark is set for the sector and where 'benchmark participants', including retailers and certain other entities such as large electricity users, are allocated a share of the sector's benchmark relative to their share of the electricity market.[58] If a benchmark participant's emissions are at or below the level of its allocated share of the benchmark then there is no scheme liability. Where a benchmark participant's emissions exceed the individual benchmark, the entity may meet this emissions liability through the surrender of abatement certificates or face a penalty.[59] Each NSW Greenhouse Abatement Certificate (NGAC)[60] represents the abatement of 1 tonne of CO_2-e emissions.[61] NGACs can be generated by activities which increase carbon stocks in an eligible forest provided certain conditions are satisfied.[62]

An accredited entity must manage a carbon sequestration pool, which is an aggregation of forests managed for sequestration.[63] This pooling mechanism allows for permanence risks to be spread across the forests in the pool and it also allows the pooling of carbon sequestration rights across a large number of smaller landowners, where the manager of the pool acts as an agent for the landowners.[64]

Consistent with the requirements under the Kyoto Protocol, an eligible forest must be planted on or after 1 January 1990 on land in NSW that is 'Kyoto-Consistent Land'.[65] The annual change in carbon stock must be estimated using a method that is consistent with the principles set out in the relevant standard.[66] The pooling mechanism under the GGAS allows for the sequestration pool manager to maintain permanent storage at an average level while allowing for harvesting across the plantations in the pool.[67] This achieves a similar result to the averaging approach proposed for harvested forests under the CPRS. Among other conditions, persons seeking accreditation must be able to demonstrate that they can maintain the greenhouse gas abatement achieved by the carbon sequestration activities for 100 years from the date the NGACs are registered.[68]

The applicant must also have appropriate risk management procedures in place in relation to the sink forest (for example, to prevent against fire, diseases and pests).[69] If the forest is physically lost through harvesting, fires or other depletion processes after the NGACs have been granted and the pool is unable to compensate for this loss, the value of the NGACs is clawed back by an equal and offsetting emission in the year of the loss.[70]

IV. FORESTRY INCENTIVES IN NEW ZEALAND

There are currently three complementary measures in New Zealand designed to encourage landowners to establish new and preserve existing forests: the forest related components of the New Zealand Emissions Trading Scheme (NZ ETS); Permanent Forest Sink Initiative (PFSI); and Afforestation Grant Scheme (AGS).

A. NZ ETS

Forestry was the first sector to enter the NZ ETS in January 2008, reflecting the New Zealand government's view that forests would play significant role in New Zealand's climate change initiatives.[71] The Kyoto delineation between pre-1990 and post-1989 forests is a key feature of the regime for forestry under the NZ ETS. Owners of post-1989 forests participate on a voluntary basis in the NZ ETS provided that land is not already subject to a forest sink covenant under the PFSI scheme.[72] Such an owner can receive New Zealand Units (NZUs) for the increase in carbon stocks in the forest from 1 January 2008.[73] Land owners have until the end of 2012 to decide whether to participate and can then received units for all carbon sequestered between 2008 and 2012.[74] A mandatory emissions return covering the commitment period (2008–2012) must be lodged in early 2013.[75] The change in carbon stocks is quantified using either a look up table or the field measurement approach.[76] A participant will be liable to surrender NZUs or other carbon units such as AAUs in the case of a decrease, such as that resulting from harvesting or natural causes,[77] up to a limit of the number of NZUs issued in relation to the relevant forest.[78]

Owners of pre-1990 forests must participate in the NZ ETS when more than 2 hectares is deforested in any five-year period beginning on 1 January 2008.[79] The owner must file an emissions return for every year in which the deforestation occurs in order to determine the liability to surrender NZUs to reflect the emissions generated by the deforestation.[80] A person ceases to be a participant upon ceasing deforestation.[81] Owners of pre-1990 forest land may be eligible for a free allocation of NZUs under the pre-1990 Forest Land Allocation Plan.[82] This plan is designed to partially compensate landowners for the reduction in the value of their land given that application of the NZ ETS.[83]

B. Permanent Forest Sink Initiative

The PFSI commenced in July 2008 and aims to promote the establishment and maintenance of permanent forests on previously unforested land.[84] Under this initiative a landowner is granted AAUs for carbon sequestered over the first commitment period (2008–2012) in a permanent forest established on or after 1 January 1990 if certain additional requirements are met[85] and on application to the Secretary for Agriculture and Forestry.[86]

There are three key requirements for acquiring AAUs.[87] First, consistent with the Kyoto Protocol, the forest must have been established on or after 1 January 1990 on land that was not used as forest land as at 31 December

1989.[88] Second, the landowner must enter into a covenant with the government for 99 years based on a forest plan and there are restrictions on when the covenant can be removed.[89] Third, a qualifying forest must only be harvested consistently with approved harvesting practice.[90] Broadly the PSFI Guidelines allow only progressive harvesting of individual trees or small coupes but require at least 80 per cent of the basal area to be maintained.[91] The number of units granted under the PFSI is based on the carbon stock assessment during the commitment period. The carbon stock assessment is made using either standard look up tables or by direct measurement.[92] If the carbon stock assessment shows that the carbon stored in the forest is below the assessed level or if there is a decrease in the carbon stock of the eligible forest, the landowner is then obliged to transfer to the Crown one AAU for each whole tonne of carbon stock decease.[93] Landowners are also required to return carbon units if they harvest their forest contrary to approved harvesting practice plus they must pay a penalty.

C.   Afforestation Grant Scheme

The AGS is a competitive grant scheme that is jointly funded by the New Zealand government and regional councils and aims to increase the area of Kyoto-compliant new forest in New Zealand.[94] The AGS is complementary to the PSFI and NZ ETS and was designed to provide a simpler option for landowners to obtain benefits from establishing a new forest compared with the other schemes.[95]

Under the AGS a landowner or a person who has a right to use land for forestry can apply for a grant to establish a new forest. The application must identify the land for the proposed new forest and the land must satisfy a number of eligibility requirements. The key requirements are that the land must: be of an area between 5 and 300 hectares in size;[96] either must not have been forest land on 31 December 1989 or was forest land on 31 December 1989 but was deforested between 1 January 1990 and 31 December 2007;[97] and not be subject to another government grant scheme including the NZ ETS and PFSI.[98] Successful applicants are required to enter into a grant agreement with the Minister of Agriculture which provides that the grantee will establish a new forest in the agreed location and maintain it for 10 years from the date of the agreement.[99] The grantee receives funding once the forest has been established and maintained to the 'minimum establishment standard'.[100] Upon completion of the 10-year agreement period, the landowner is free to do what it wishes with the forest, including transfer to the PFSI or the ETS or harvesting.[101]

The grant agreement provides that the grantee owns the established forest and can generate income from timber harvesting provided that minimum tree coverage is maintained during the 10-year term[102] and the land is not deforested.[103] The Crown owns any carbon credits that the forest generates and can use them to meet any liabilities for harvesting and deforestation in relation to the forest under the Kyoto Protocol.[104] The grantee must also take steps to protect the forest from damage or destruction.[105] If the forest is destroyed by natural causes and the grantee took the appropriate steps to protect against those causes then there is no liability on the part of the landowner and the Crown accepts the risk of carbon loss.[106]

V. COMPARING THE APPROACHES

A comparison of the incentives linked to forest carbon sinks in Australia and New Zealand reveals significant shortcomings in the Australian scheme. As noted above, the various incentives in place should work coherently to increase the comparative financial advantage of using land as a carbon sink rather than for other purposes such as plantation forestry or agriculture. The Australian scheme clearly falls short given that the only nationally available incentive is a concession for establishment expenditure where the after-tax cost of establishment is still borne by the taxpayer and there is no federal mechanism to produce income returns from the carbon sink. It is also problematic that the tax concession only provides for an upfront assessment and there is no clear commitment or monitoring mechanism. It may well be that this mechanism has been put in place in contemplation that it would be coordinated with the CPRS. However, given the continued uncertainty regarding that scheme it would be preferable if the conditions for the concession were tightened. Several features of the proposals under the CPRS for reforestation are of interest, including the averaging mechanism for harvested forests and the 'risk of reversal buffer' for permanent forests, but, for the time being, the only available mechanism to produce the requisite return is the GGAS in NSW.

The GGAS provides a useful model and contains some notable advantages such as the pooling element and the 100-year commitment to ensure permanence. The pooling mechanism offers many advantages, including the spreading of risks of carbon loss, the ability to average over forests harvested on a rotational basis and the option for smaller landholders to access the scheme through a pool manager. One issue not addressed by the GGAS that should be considered by the Australian government is the inclusion of a mechanism to provide an incentive to preserve pre-existing forests from deforestation.

The set of mechanisms in place in New Zealand reflects a more thorough consideration of the issues and limits of using different forms of incentives to encourage carbon sink forestry. The structure of the AGS reflects the fact that is it designed to cater for holders of smaller parcels of land where these landholders might have otherwise found the compliance requirements for the PFSI and the NZ ETS too burdensome. The relative simplicity of the AGS is shown by the upfront grant, the 10 year commitment period and the risk limitation if the forest is lost due to natural causes.

In contrast, the PFSI may be more appropriate for larger scale forestry operators who can commit to the requirements of the covenant required by the scheme and assume the 99-year limit on large-scale harvesting. This incentive relies on year-to-year measurement of carbon stores to support the issue of AAUs but does allow for small-scale harvesting. Additional risk is assumed by the operator under the clawback mechanism and the penalty for large-scale harvesting. The addition of the NZ ETS mechanisms allows a complementary scheme for reforestation but also extends emissions liabilities to deforestation of pre-1990 forest land, which is an important feature.

VI. CONCLUSIONS AND CHALLENGES

This analysis of mechanisms currently in place or proposed for Australia and New Zealand highlights some of the challenges presented in the design of carbon sink forestry incentives. Ideally, mechanisms should be in place to address both the emissions produced from deforestation and the carbon sequestration opportunities provided by reforestation so that there are incentives for preserving existing forests and establishing new forests, particularly in areas that may not be best suited for other land uses such as food production. The scheme design must address the issue of permanence through long-term commitment periods and should ideally include mechanisms whereby holders of both large and small land parcels have an incentive to participate, with the commitment period and risk management issues perhaps warranting modification for smaller holders. From a commercial and environmental point of view, limited harvesting should be considered. This would provide an additional financial return to forestry operators and also would acknowledge that the carbon emissions from limited harvesting can, in a relatively short period, be offset by the additional carbon absorption provided by new growth.

Forest carbon sinks offer real opportunities as one component of a comprehensive strategy to address the rising level of atmospheric carbon, but it must be acknowledged that there are limits to its potential contribution. Reducing emissions from deforestation can produce an immediate impact on global emissions but forest-based carbon sequestration produces benefits over the longer term given the relatively slow growth rates of trees. It must also be acknowledged that any particular jurisdiction will have a limit to the amount of land suitable for carbon sink establishment, given other important land uses, such as food production. There are also many scientific issues to be further considered, including the suggestion that on maturity a forest will cease to continue to absorb carbon and may in fact become a net emitter.[107]

NOTES

1. This research was supported by a Discovery Project Grant provided by the Australian Research Council (DP1096061). Research assistance was provided by Ms Alex Evans.
2. It is estimated that the forestry sector contributed around 17.4% of anthropogenic greenhouse gas emissions in 2004: Intergovernmental Panel on Climate Change (IPCC) (2008), *Climate Change 2007: Synthesis Report* (Fourth Assessment Report) 36.
3. *Ibid.* 60, 68.
4. For further information on REDD, see Lyster R. (2009), 'The new frontiers of climate law: reducing emissions from deforestation (and degradation)', *Environmental Planning & Law Journal* **26**, 417. Growing international support for REDD was confirmed at the Fifteenth Conference of the Parties to the UNFCCC (COP 15) and work to establish a formal mechanism is continuing: COP 15, Decision 4/CP.15 (18–19 December 2009), http://unfccc.int/documentation/decisions/items/3597.php?such= j&volltext=/CP.15#beg.
5. See also Gould, Karen, Monique Miller and Martijn Wilder, 'Legislative Approaches to Forest Sinks in Australia and New Zealand: Working Models for Other Jurisdictions?' in Streck, Charlotte, Robert O'Sullivan, Toby Janson-Smith, *et al.* (eds) (2008), *Climate Change and Forests: Emerging Policy and Market Opportunities*, London, UK: Chatham House and Washington DC, US: Brookings Institution Press, 253–71, provides a useful description of many of these mechanisms (some of which have now been enacted as law) at an earlier stage in their development.
6. Streck *et al.* (2008), 'Climate Change and Forestry: An Introduction' in Streck *et al.* (eds) (n 5 above) 4.
7. It is difficult to estimate the carbon currently stored in forests. This is reflected in the variation in figures quoted. Nijnik states that existing forests hold 120Gt of carbon: Nijnik, Maria (2010), 'Carbon Capture and Storage in Forests' in Hester, Ronald E. and Roy M. Harrison (eds) (2010), *Carbon Capture: Sequestration and Storage*, Cambridge, UK: The Royal Society of Chemistry Publishing, 206. In comparison, Streck quotes a figure of 283 Gt stored in forest biomass and 638 Gt stored in forest ecosystems: Streck *et al.* (2008) (n 6 above) 4. See also Corbera E., M. Estrada and K. Brown (2010), 'Reducing greenhouse gas emissions from deforestation and forest degradation in developing countries: revisiting the assumptions', *Climatic Change* **100**, 355, 358.
8. Nijnik (2010) (n 7 above) 206.
9. Chapman, Stephen J. (2010), 'Carbon Sequestration in Soils' in Hester and Harrison (eds) (n 7 above) 179–202.

10. Turley, C., J. Blackford, N. Hardman-Mountford, *et al.* (2010), 'Carbon Uptake, Transport and Storage by Oceans and the Consequences of Change' in Hester and Harrison (eds) (n 7 above) 240–84.

11. Nijnik (2010) (n 7 above) 208.

12. *Ibid.*

13. *Ibid.* 209.

14. *Ibid.* 206.

15. *Ibid.*

16. For a discussion of the permanence issue, see Ebeling, Johannes (2008), 'Risks and Criticism of Forestry-Based Climate Change Mitigation and Carbon Trading' in Streck *et al.* (eds) (n 5 above) 43–58.

17. Streck *et al.* (2008) (n 6 above) 5.

18. *Kyoto Protocol to the Framework Convention on Climate Change* 37 ILM 22 (1998); Lyster (2009) (n 4 above) 420.

19. Streck *et al.* (2008) (n 6 above) 3.

20. *Ibid.* 3–6; Lyster (2009) (n 4 above); Ebeling (2008) (n 16 above) 43–58.

21. See, e.g., McKinsey & Co, *Pathways to a Low Carbon Economy* (2009), Summary of Findings 9–11; Stern N. (2007), *The Stern Review: The Economics of Climate Change*, UK: Cambridge University Press, Ch 25.

22. EcoSecurities (2010), *The Forest Carbon Offsetting Report 2010,* www. ecosecurities.com/Standalone/Forest_carbon_offsetting_report_2010/default.aspx.

23. UNFCCC, Project design document for afforestation and reforestation project activities (CDM-A/R-PDD), http://cdm.unfccc.int/Projects/pac/pac_ar.html. For guidelines specific to forestry CDM projects see UNFCCC, Guidelines, Guidance and Clarifications, http://cdm.unfccc.int/Reference/Guidclarif/index.html.

24. UNFCCC, CDM Project Registry, http://cdm.unfccc.int/Projects/registered.html.

25. EcoSecurities (2010) (n 22 above) 9.

26. *Ibid.* 5.

27. Enters T., P. Durst and C. Brown (2003), 'What Does It Take? The Role of Incentives in Forest Plantation Development in the Asia-Pacific Region' (24–30 March 2003), UNFF Intersessional Experts Meeting on the Role of Planted Forests in Sustainable Forest Management, New Zealand, 3.

28. *Ibid.*

29. *Ibid.* 3–4.

30. *Ibid.*

31. Australian Government, Department of Climate Change and Energy Efficiency, *National Carbon Offset* (effective 1 July 2010) 3–5.

32. *Electricity Supply Act 1995* (NSW), s 97F; *Electricity Supply (General) Regulation* 2001 (NSW), cl 73GA; *Greenhouse Gas Benchmark Rule (Carbon Sequestration)* No 5 of 2003 (NSW) (NSW Carbon Sequestration Rule).

33. Australia, Federal Parliament, Senate Rural and Regional Affairs and Transport References Committee, *A Review of Plantations for Australia: The 2020 Vision* (2004), Chapter 2, www.aph.gov.au/Senate/committee/rrat_ctte/completed_inquiries/2002–04/plantation_forests/report/index.htm.

34. *Income Tax Assessment Act 1936* (Cth) (ITAA 1936), s 82KME, allowed prepaid forestry expenditure to be deducted up front; *Income Tax Assessment Act 1997* (Cth) (ITAA 1997), Division 394, has applied since 2007.

35. ITAA 1997, Subdiv 40-J.

36. ITAA 1997, ss 40–1010 and 40–1015.

37. Explanatory Memorandum to *Tax Laws Amendment Act (2008 Measures No 2) 2008* (Cth), paras 16.33–16.41.

38. ATO Interpretive Decision 2009/60 'Capital allowances: carbon sink forest – cost of land', http://law.ato.gov.au/.

39. ITAA 1997, s 40–1010(2).
40. ATO, 'Carbon sink forests', www.ato.gov.au/taxprofessionals/content.asp?doc=/content/00103282.htm&page=1&H1.
41. ATO, 'Notice of establishment of trees in a carbon sink forest', www.ato.gov.au/content/downloads/atp00103282carbonsink.pdf; 'Environmental and Natural Resource Guidelines relating to managing trees for carbon storage', www.comlaw.gov.au/ComLaw/Legislation/LegislativeInstrument1.nsf/0/E7FEEEA983FCFAF4CA257513000FDE9A?OpenDocument.
42. ITAA 1997, ss 40–1010(2), (5) and (6).
43. The value of tax revenue forgone due to this measure has been valued at $5m for the 2008/09 income year and is predicted to peak at $13m in the 2011/12 year: Australian Government, Treasury, *Tax Expenditure Statement 2009*, 108.
44. *Carbon Pollution Reduction Scheme Bill* 2010 (Cth) (CPRS Bill), s 191.
45. Australian Government, DCCEE (2008), *Carbon Pollution Reduction Scheme: Australia's Low Pollution Future* (White Paper), Ch 6, policy position 6.27.
46. Kevin Rudd, then Prime Minister (27 April 2010), 'Transcript of Doorstop at Nepean Hospital, Penrith', http://pmrudd.archive.dpmc.gov.au/node/6708; Australian Government, DCCEE (5 May 2010), 'Carbon Pollution Reduction Scheme', www.climatechange.gov.au/en/media/whats-new/cprs-delayed.aspx.
47. Emissions Trading Division, DCCEE (2009), Carbon Pollution Reduction Scheme Stakeholder Consultation, *Discussion Paper: Design Issues relating to Reforestation* (Discussion Paper), www.climatechange.gov.au/government/initiatives/cprs/who-affected/reforestation.aspx.
48. *CPRS Bill*, Part 1, s 5, 'project area' definition; Part 10, Division 5.
49. *Ibid.*, 'forest stand' definition.
50. *Discussion Paper* (n 47 above) para 10.
51. *White Paper* (n 45 above) 54–5.
52. *CPRS Bill*, s 220.
53. *Discussion Paper*, (n 47 above) paras 25–8; *White Paper*, (n 45 above) 55–8.
54. *Discussion Paper*, (n 47 above) para 29; *White Paper*, (n 45 above) 55–8; *CPRS Bill*, Part 10, Divisions 6, 7 and 8.
55. *CPRS Bill*, s 217.
56. *Ibid.* s 232.
57. New Part 8A was inserted in *Electricity Supply Act 1995* (NSW) by *Electricity Supply Amendment (Greenhouse Gas Emission Reduction) Act 2002* (NSW).
58. 'Introduction to the Greenhouse Gas Reduction Scheme', 3, www.greenhousegas.nsw.gov.au/documents/Intro-GGAS.pdf. Carbon dioxide equivalent of greenhouse gas emissions is a measure of equivalent global warming potential, where greenhouse gas means carbon dioxide, methane, nitrous oxide, a perfluorocarbon gas and sulphur hexafluoride. *Electricity Supply Act* 1995 (NSW), s 97AB; *Electricity Supply (General) Regulation* 2001, cl 73A.
59. NSW GGAS website, www.greenhousegas.nsw.gov.au. A liability may also be met through the surrender of Renewable Energy Certificates issued under the national Mandatory Renewable Energy Target: *Renewable Energy (Electricity) Act 2000* (Cth).
60. *NSW Carbon Sequestration Rule*, rule 11 'Definitions and Interpretation'.
61. *Electricity Supply Act 1995* (NSW), s 97EA.
62. *Ibid.* s 97 EB; *Electricity Supply (General) Regulation* 2001 (NSW), cl 73GA; *NSW Carbon Sequestration Rule*.
63. Gould *et al.* (n 5 above) 261.
64. *Ibid.* 262–3.
65. *NSW Carbon Sequestration Rule,* rule 11, definitions for 'Eligible Forest' and 'Kyoto-Consistent Land'.

66. Currently AS 4978.1–2006 *Quantification, Monitoring and Reporting of Greenhouse Gases in Forestry Projects: Part 1 – Afforestation and Reforestation*, section 3.22: *NSW Carbon Sequestration Rule*, rule 8.

67. NSW GGAS website, 'Carbon sequestration – forestry', www.greenhousegas.nsw.gov.au/acp/forestry.asp.

68. *NSW Carbon Sequestration Rule*, rule 5(c).

69. *Ibid.* rule 5(d).

70. *Ibid.* rule 7.3.

71. New Zealand, Ministry of Agriculture and Forestry, (2010), *A Guide to Forestry in the Emissions Trading Scheme*, 7.

72. *Climate Change Response Act* 2002 (NZ) (CCR Act), s 57 and Sch 4, Part 1; *ibid.* Ch 2.

73. *CCR Act*, s 64.

74. *A Guide to Forestry in the Emissions Trading Scheme*, (n 71 above) 8.

75. *Ibid.* 18.

76. *Ibid.* 13.

77. *Ibid.* 18–19.

78. *Ibid.* 19.

79. *CCR Act*, s 57 and Sch 3, Part 1; *A Guide to Forestry in the Emissions Trading Scheme*, (n 71 above) Ch 3.

80. *CCR Act*, s 63.

81. *A Guide to Forestry in the Emissions Trading Scheme*, (n 71 above) 25.

82. *CCR Act*, ss 70 and 72.

83. *A Guide to Forestry in the Emissions Trading Scheme*, (n 71 above) 26.

84. *Forests Act 1949* (NZ), Part 3B.

85. NZ, Ministry of Agriculture and Forestry (MAF), *Permanent Forest Sink Initiative Guidelines* (2008) (PFSI Guidelines), 3.

86. Forests (Permanent Forest Sink) Regulations 2007 (NZ) (SR 2007/354) (PFS Regulations), regulations 10(1) and (2); Forest Sink Covenant (2009), www.maf.govt.nz/forestry/pfsi/#Outline.

87. *PFSI Guidelines* (n 85 above) 5–13.

88. 'Kyoto-compliant land' definition for the purposes of PFS Regulations, regulation 3.

89. PFS Regulations, regulations 3 ('restricted period' means 99 years from registration of the covenant) 4 and 6.

90. *Ibid.* regulations 3 ('approved harvesting practice' definition) and 7.

91. *Ibid.* regulation 3 ('continuous cover forestry' definition).

92. NZ, MAF has prepared a report setting out a proposed PFSI carbon accounting system: PFSI Carbon Accounting Design Team (2007), *Proposed PFSI Carbon Accounting System*, www.maf.govt.nz/forestry/pfsi/#Outline.

93. Forest Sink Covenant, (n 86 above) Sch 4.

94. NZ, MAF (2010), *A Guide to the Afforestation Grant Scheme* (AGS Guide) 3.

95. *Ibid.*

96. *Ibid.* 5.

97. Land can also be eligible if it was forest land on 31 December 1989, was forested after 31 December 2007 (when the NZ ETS took effect) and any deforestation liabilities under the ETS have been satisfied: *ibid.* 5.

98. *Ibid.* 7.

99. Template AGS Grant Agreement, www.maf.govt.nz/climatechange/forestry/initiatives/ags/.

100. Ibid clause 2 ('Minimum Establishment Standard' definition); *AGS Guide*, (n 94 above) 7.

101. *AGS Guide*, (n 94 above) 4.

102. AGS Grant Agreement, (n 99 above) clause 5.2 requires the landowner to maintain the minimum establishment standard for four years from achievement of this standard plus maintain at least 500 stems per hectare for the balance of the term.

103. *Ibid.* clause 6.
104. *Ibid.* clause 11.
105. *Ibid.* clause 5.3.
106. *AGS Guide*, (n 94 above) 11.
107. Nijnik (2010), (n 7 above) 207.

PART V

Other Environmental Taxation Schemes

12. Price signal or tax signal? An international panel data analysis on gasoline demand reaction

Seung-Joon Park

1. INTRODUCTION

In general, the effect of energy taxes on energy consumption is estimated by calculating the price elasticity, where the tax rate is treated merely as a part of consumer prices. However, it has been argued that a unit change in tax rate has a stronger effect on energy demand than a similar change in price because of the so called 'signaling effect of tax' (Ghalwash 2007).

The argument for the 'signaling effect' follows. Once an energy tax is set, it is likely to be maintained for a long time. On the other hand, energy demand responds not only to current price changes but also to future price prospects. In particular, energy consumers choose their appliances based on the future outlook of price development. Therefore, the tax rate, which is unlikely to decrease again, will have a stronger effect on demand than prices do, as prices may unexpectedly decrease sooner or later. Although it is commonly known that the long-term price elasticity is generally greater than short-term elasticity, the signaling effect may be exerted soon after a change in the tax rate.

The empirical studies related to this effect are, for example, Ghalwash (2007) and Bardazzi, Oropallo and Pazienza (2009). Both studies proved that 'tax elasticity' is greater than 'price elasticity'. The former is based on Swedish consumption data using the Almost Ideal Demand System (AIDS) and the latter is based on Panel Data analysis with data from 5600 Italian companies.

However, the 'tax elasticity' calculated using log-linear specification is difficult to interpret and may be biased. We, therefore, support our estimation of 'tax elasticity' by using another estimation method with a simple linear specification. We used the OECD's long-term statistics on gasoline price, taxes as well as consumption and applied the Panel Data methodology. In our study, the emphasis is not on the absolute amount of the effect, but rather on the method to evaluate the difference between the price and tax effects.

2.   SPECIFICATION OF THE TAX EFFECT

If we assume that the effects of a unit change in the pre-tax price (1 cent/litre) and the excise tax rate (1 cent/litre) are identical, then we can easily calculate the (after-tax) price elasticity of gasoline demand using regression analysis with log-linear specification.

$$\ln Y = \beta_0 + \beta_1 \ln X + \beta_2 \ln (\overline{p} + t) \tag{1}$$

where Y is gasoline demand, X is real GDP, $\overline{p}$ is the pre-tax price of gasoline (cent/litre) and t is the gasoline tax rate (specific tax, cent/litre). Then β_0 is constant, β_1 is GDP elasticity and β_2 stands for price elasticity. However, if we need to differentiate the effect of the pre-tax price and the tax rate, separating the third term of the right side of eq. (1) may be confusing.[1] An inaccurate separation may lead to misinterpretations. Therefore, Ghalwash (2007), for example, defined the 'tax index (τ)' as unit plus 'ad valorem tax rate', although, in reality, most energy taxes are specific taxes [cent/litre]. He introduced the following expressions:

$$p = \overline{p} \cdot \tau \tag{2}$$

$$\ln \overline{p} = \ln \overline{p} + \ln \tau \tag{3}$$

where p is the price including taxes, $\overline{p}$ is the price without taxes and τ is the 'tax index' corresponding to $(\overline{p} + t)/\overline{p}$ or, as easily understood, $\tau = p \div \overline{p}$. That is, if the ad valorem tax rate is 20%, the tax index is 1.2. A specific tax rate can easily be converted to an ad valorem tax rate; therefore, $\tau = 1.5$ if the pre-tax price of gasoline is 1 [$/litre] and the specific tax rate is 50 [cent/litre]. In this manner, the change in prices and taxes can be treated similarly. Then, we can modify eq. (1) using eq. (3) and allow for the difference between 'price elasticity (β_2)' and 'tax elasticity (β_3)' as follows.

$$\ln Y = \beta_0 + \beta_1 \ln X + \beta_2 \ln \overline{p} + \beta_3 \ln \tau \tag{4}$$

We primarily follow this method. However, we should keep in mind that this specification may be difficult to interpret and the estimated coefficient may also be biased, because (1) the tax index changes if the pre-tax price changes while the specific tax rate remains unchanged, and (2) for consumers, a unit increase in the specific tax is not easy to convert to τ.

Let us consider the following examples:

(1)   If the pre-tax price increases from 100 [cent/litre] to 200 [cent/litre] while the specific tax rate remains fixed at 50 [cent/litre], τ will be *reduced* from 1.5 to 1.25.

(2)   Assume that the pre-tax price and specific tax rate are 50 [cent/litre] initially. Then, if both price and tax rate increase by 10 [cent/litre], τ remains unchanged ($\tau = 2$) before and after the change, even though the tax rate has indeed been increased (before: $\tau = (50 + 50)/50$, after: $\tau = (60 + 60)/60$).

These examples clearly indicate the difficulty in interpreting τ.

On the other hand, Bardazzi, Oropallo and Pazienza (2009) expressed their specification as:

$$\ln Y = \beta_0 + \beta_1 \ln X + \beta_2 \ln \bar{p} + \beta_3 \ln tax \qquad (5)$$

where a *tax* is referred to as a 'tax component'. As no detailed explanation of *tax* has been provided, it may be interpreted as (1) the tax index of Ghalwash (2007), (2) a specific tax rate, or (3) a specific tax rate converted to an ad valorem tax rate. If it is interpreted as (1), the same argument applies. The problem that accompanies (2) and (3) is identical under a log-linear specification. Although a unit tax hike, say by 1 [cent/litre], will have the same effect on demand regardless of initial tax level, the logarithm of the unit tax hike has greater value as the initial tax level decreases. This problem remains unchanged even if the specific tax rate is converted to an ad valorem tax rate [%].

Therefore, we complement the specification of existing studies with a simple 'linear' specification:

$$Y = \beta_0 + \beta_1 X + \beta_2 \bar{p} + \beta_3 t \qquad (5)$$

In this case, the pre-tax price ($\bar{p}$) and the specific tax (t) have the same unit [cent/litre], making it very easy to compare the price effect (β_2) and the tax effect (β_3). Nevertheless, since we cannot know whether the difference is significant, we have to develop another equation. The simple linear eq. (5) can be transformed:

$$Y = \beta_0 + \beta_1 X + \beta_2 \bar{p} + \beta_2 t - \beta_2 t + \beta_3 t$$

$$Y = \beta_0 + \beta_1 X + \beta_2 (\overline{p} + t) + (\beta_3 - \beta_2)t \qquad (6)$$

As price coefficients must be negative ($\beta_2 < 0$, $\beta_3 < 0$), if tax effect (β_3) is stronger than the price effect (β_2), then β_3 is smaller than β_2, so $\beta_3 - \beta_2$ must be significantly negative.

In the next section, we describe the data used and how the difference was estimated.

3. DATA AND ESTIMATION

We used the databases *Energy Prices & Taxes – Quarterly Statistics* and *Energy Balances of OECD Countries* from the International Energy Agency (IEA). Price data are from 1978 to 2007 and gasoline consumption and related economic data are from 1960 to 2006 as Panel for 29 OECD countries.[2] But there is a non-negligible number of missing values. In particular, tax rate values were often missing, resulting in only 746 observations for 29 countries.

On the other hand, these databases enable us to make an international comparison with several dollar-based economic indicators. As differences among prices and tax rates exist between 'Premium' and 'Regular' or between 'Leaded' and 'Unleaded' and for most countries we do not have a long continuous time series for any category of gasoline (because, for example, leaded gasoline is no longer used), we carefully combined the time series of different fuel categories to create as many available observations as possible (see the Appendix on page 181). We converted economic values including prices and tax rates into dollar-based values by using purchasing power parity (ppp), which was again deflated by the U.S. consumer price index.

The equations for the log-linear and linear estimation are as follows:

$$\ln gaspc_{it} = \alpha_i + \beta_1 \ln gdppc_{it} + \beta_2 (\ln pr_{it} + \ln tfact_{it}) + \qquad$$
$$(\beta_3 - \beta_2)tfact_{it} + \mu_{it} \qquad (7)$$

$$gaspc_{it} = \alpha_i + \beta_1 gdppc_{it} + \beta_2 (pr_{it} + tr_{it}) + (\beta_3 - \beta_2)tr_{it} + \mu_{it} \qquad (8)$$

where $gaspc_{it}$ is the per capita gasoline consumption [kg/person], $gdppc_{it}$ is the per capita real GDP [1000\$ppp/person], pr_{it} is the real pre-tax gasoline price [\$ppp/litre], tr_{it} is the real excise tax rate on gasoline [\$ppp/litre], $tfact_{it}$ is the 'tax factor' (($pr_{it} + tr_{it})/pr_{it}$), μ_{it} is the error term and suffix i denotes the country number while suffix t denotes the year. The base year for real prices is 2000.

Many countries have a value-added tax (VAT) or sales tax, both of which are ad valorem taxes (henceforth VAT). VAT will boost the change in the pre-tax price and the specific tax rate, respectively, because $(1 + T)(\bar{p} + t) = (1 + T)\bar{p} + (1 + T)t$, where T is the VAT rate. Therefore, in our definition, tr_{it} includes the VAT added to the excise tax rate (that is, $tr_{it} = (1 + T)t$), while pr_{it} consists of the pre-tax price and VAT (that is, $pr_{it} = (1 + T)\bar{p}$).

Basic statistics on historical tax rates for gasoline (tr_{it}) in 29 economies are shown in Table 12.1 (nominal value in ppp dollar). As is well known, the tax rate is extremely low in the US and relatively low in Japan.

Table 12.1 Tax rate on gasoline (tr_{it}) [PPP$/litre]

Country	N	mean	s.d.	min.	max.
Australia	31	0.23102	0.10997	0.05000	0.35385
Austria	31	0.49531	0.14726	0.23760	0.72000
Belgium	31	0.55664	0.19152	0.22040	0.80667
Canada	31	0.16074	0.07531	0.02838	0.23963
Czech Republic	19	1.05982	0.26131	0.84439	10.65741
Denmark	31	0.47367	0.13224	0.20291	0.64608
Finland	31	0.51672	0.25631	2.00000	0.85400
France	31	0.61983	0.20355	0.27930	0.87707
Germany	31	0.54849	0.25426	0.20533	0.93878
Greece	31	0.56547	0.13544	0.31733	0.82600
Hungary	29	0.96196	0.26421	0.14234	1.27575
Ireland	31	0.53155	0.12381	0.22000	0.66550
Italy	31	0.75129	0.11137	0.44800	0.87000
Japan	31	0.32337	0.07948	0.17234	0.48819
Korea	9	1.05540	0.03080	1.00568	1.09215
Luxembourg	29	0.41046	0.14493	0.16333	0.65167
Mexico	26	0.12136	0.16139	0.00000	0.43405
Netherlands	31	0.60850	0.23945	0.19969	0.92556
New Zealand	31	0.25964	0.06903	0.15556	0.39000
Norway	31	0.48223	0.20968	0.16709	0.72917

Country	N	mean	s.d.	min.	max.
Poland	15	0.87648	0.23694	0.51850	1.14295
Portugal	31	0.86560	0.14467	0.48471	1.15000
Slovak Republic	25	0.94936	0.16689	0.63960	1.50000
Spain	31	0.52815	0.15821	0.12500	0.71257
Sweden	31	0.50699	0.21536	0.15484	0.81731
Switzerland	31	0.38496	0.10191	0.23320	0.55145
Turkey	12	1.44900	0.34478	0.80500	2.03550
United Kingdom	31	0.61831	0.27274	0.19600	0.95958
United States	31	0.08677	0.02638	0.03000	0.13000
Total	815	0.54041	0.31892	0.00000	2.03550

Scatter diagrams of these variables are shown in Figures 12.1, 12.2 and 12.3. These charts show that almost no correlation exists between *gdppc* and *gaspc* within a country and that the correlation between *gaspc* and *tr* (tax rate) is greater than between *gaspc* and *prtr* (after-tax price, *pr + tr*).

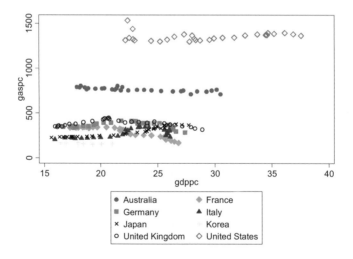

Figure 12.1 Correlation between gaspc and gdppc

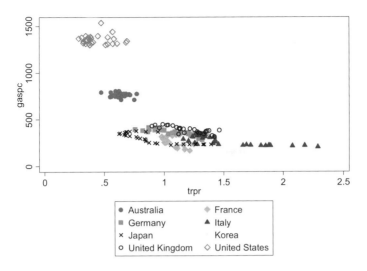

Figure 12.2 Correlation between gaspc and trpr

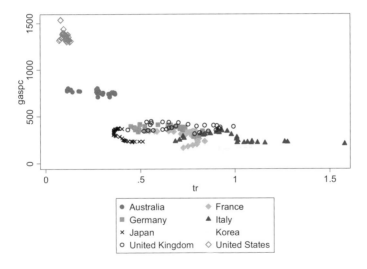

Figure 12.3 Correlation between gaspc and tr

4. ESTIMATION

4.1. Test of endogeneity

The independent variables in eq. (7) are considered to be random variables because of the possibility of endogeneity. In particular, the after-tax gasoline price in each country may depend on the relationship between supply and demand. Therefore, we apply the endogeneity test following Wooldridge (2006, p. 527). We show eq. (7) and eq. (8) here again.

$$\ln gaspc_{it} = \alpha_i + \beta_1 \ln gdppc_{it} + \beta_2 (\ln pr_{it} + \ln tfact_{it}) + (\beta_3 - \beta_2)tfact_{it} + \mu_{it} \tag{7}$$

$$gaspc_{it} = \alpha_i + \beta_1 gdppc_{it} + \beta_2 (pr_{it} + tr_{it}) + (\beta_3 - \beta_2)tr_{it} + \mu_{it} \tag{8}$$

The tax rate tr_{it} is assumed to be exogenous. We also assume that $tfact_{it}$ is exogenous, although it may be disputable. As $pr_{it} + tfact_{it}$ and $pr_{it} + tr_{it}$ are considered as endogenous variables, we apply real imported crude oil price $pcor_{it}$ [US$/barrel] as an instrument variable. Although the per capita GDP ($gdppc_{it}$) may also be an endogenous variable, we could not find any suitable instrument variable, so we assume that it is exogenous.

$$\ln(pr_{it} + tfact_{it}) = \chi_i + \delta_1 \ln gdppc_{it} + \delta_2 \ln tfact_{it} + \delta_3 \ln pcor_{it} + \hat{v}_{it}^1 \tag{9}$$

$$pr_{it} + tr_{it} = \chi_i + \delta_1 dgppc_{it} + \delta_2 tr_{it} + \delta_3 pcor_{it} + \hat{v}_{it}^2 \tag{10}$$

Residuals $\hat{v}_{it}^1$ and $\hat{v}_{it}^2$ are calculated by estimating eq. (9) and eq. (10). Based on eq. (7) and eq. (8), by including $\hat{v}_{it}^1$ and $\hat{v}_{it}^2$ as explanatory variables, we obtain:

$$\ln gaspc_{it} = \alpha_i + \beta_1 \ln gdppc_{it} + \beta_2 (\ln pr_{it} + \ln tfact_{it}) + (\beta_3 - \beta_2) \ln tfact_{it} + \beta_4 \hat{v}_{it}^1 + \varepsilon_{it} \tag{11}$$

$$gaspc_t = \alpha_i + \beta_1 gdppc_{it} + \beta_2 (pr_{it} + tr_{it}) + (\beta_3 - \beta_2)tr_{it} + \beta_4 \hat{v}_{it}^2 + \varepsilon_{it} \tag{12}$$

If $pr_{it} + tfact_{it}$ or $pr_{it} + tr_{it}$ is exogenous, the coefficient β_4 will not be significantly different from zero.

The result of the panel data estimation of eq. (11) is as follows:

Fixed Effect Model (FE)

$$\ln gaspc_{it} = \alpha_i + 0.0325 \ln gdppc_{it} - 0.6114(\ln pr_{it} + \ln tfact_{it})$$

$$(0.446) \qquad (0.000)$$

$$-0.0604 \ln tfact_{it} + 0.1069\hat{v}_{it}^1 + \mu_{it} \qquad (13)$$

$$(0.027) \qquad (0.020)$$

Random Effect Model (FE)

$$\ln gaspc_{it} = \alpha_i + 0.0733 \ln gdppc_{it} - 0.6523(\ln pr_{it} + \ln tfact_{it})$$

$$(0.098) \qquad (0.000)$$

$$-0.1021 \ln tfact_{it} + 0.0905\hat{v}_{it}^1 + \mu_{it} \qquad (14)$$

$$(0.000) \qquad (0.060)$$

where the values in parentheses are P-values. In this case, we can conclude that $pr_{it} + tfact_{it}$ is endogenous because the coefficient of $\hat{v}_{it}^1$ is significantly different from zero (with 5% significance in FE and 10% significance in RE). The Hausman statistics for these two estimations are negative (-245.53). Therefore, we cannot decide which method is better[3].

The result of the panel data estimation of eq. (12) is as follows:

Fixed Effect Model (FE)

$$gaspc_t = \alpha_i + 0.538gdppc_{it} - 95.79(pr_{it} + tr_{it}) - 21.86tr_{it}$$
$$+ 25.24\hat{v}_{it}^2 + \varepsilon_{it} \qquad (15)$$

$$(0.261) \qquad (0.000) \qquad (0.168) \qquad (0.146)$$

Random Effect Model (FE)

$$gaspc_t = \alpha_i + 0.608gdppc_{it} - 95.72(pr_{it} + tr_{it}) - 26.83tr_{it}$$
$$+ 24.86\hat{v}_{it}^2 + \varepsilon_{it} \qquad (16)$$

$$(0.213) \qquad (0.000) \qquad (0.097) \qquad (0.160)$$

In this case, the Hausman statistics for these two estimations are negative (-6.99), so we cannot decide which estimation is better. Nevertheless, the coefficient of $\hat{v}_{it}$ is not significant at 10% level of significance so that the possibility of endogeneity is low. However, as the P-value is not sufficiently high, we will deal with endogeneity by applying the instrumental variable method (IV).

4.2. Result of the log-linear specification

The results of panel data estimation of eq. (7) are shown in Table 12.2. As the Hausman statistics for the 'Normal method' has a negative value, we cannot decide which model is better. However, the difference in the result is negligible. When we apply the Instrumental Variable method (IV) to consider endogeneity, the Hausman statistics are significant; therefore, we choose the Fixed Effect Model (FE). Using the Instrumental Variable method (IV), the number of available observations and groups are reduced (only 24 economies are available), and the estimated coefficient of per capita GDP β_1 is no more significant.

Table 12.2 Estimated results of the log-linear specification eq. (7)

	Normal method		IV method	
	Fixed	Random	Fixed	Random
ln *gdppc*	0.1567***	0.2132***	0.0501	0.0880**
(β_1)	(0.000)	(0.000)	(0.198)	(0.031)
ln *pr* + ln *tfact*	−0.5319***	−0.5763***	−0.6734***	−0.7025***
(β_2)	(0.000)	(0.000)	(0.000)	(0.000)
ln *tfact*	−0.0917***	−0.1295***	−0.1077***	−0.1406***
($\beta_3-\beta_2$)	(0.000)	(0.000)	(0.000)	(0.000)
_cons	5.4560***	5.2585***	5.8297***	5.6858***
	(0.000)	(0.000)	(0.000)	(0.000)
R^2 within	0.4264	0.4246	0.4253	0.4363
R^2 between	0.8722	0.8827	0.8637	0.8711
R^2 overall	0.7829	0.7946	0.7926	0.7996
Number of …	obs. 746, groups 29		obs. 571, groups 24	
IV			instrumented: ln *pr* + ln *tfact* instruments: ln *gdppc* ln *tfact* ln*pcor*	
Hausman Stat.	−212.00		44.35 (p = 0.000)***	

Note: P-value of t-test is in parentheses; level of significance, *** 1%, ** 5%, * 10%.
Breusch-Pagan test revealed significance at 1% level, implying that random effect panel is better than pooled OLS.

The coefficient of ln*tfact* ($\beta_3 - \beta_2$) stands for the difference between 'tax-factor' elasticity and price elasticity[4]. We should select the FE model using the IV method, as argued above and therefore, we can easily understand that the difference in elasticity is significant (−0.1077, p = 0.000) or, in other words, the tax-factor elasticity is 16.0% greater than the price elasticity ($\beta_3 = -0.7811$, $\beta_2 = -0.6734$). In this manner, we achieve results in line with Ghalwash (2007) or Bardazzi *et al.* (2009).

4.3. Results of the Linear Specification

Here, we discuss the results of eq. (8), as shown in Table 12.3. We cannot choose the best estimation from the four different methods, because the result of the endogeneity test was not significant and the Hausman statistics are negative. But the difference in results between FE and RE is negligible. The difference between the Normal method and the Instrumental Variable method (IV) is non-negligible. Using the IV method, the number of available observations and groups were reduced (only 24 economies are available) and the estimated coefficient (β_1) is no more significant.

The coefficient of *tr*, $\beta_3 - \beta_2$, is always negative and consistent with the result obtained in Section 4.2, but the gap is not significant at the 10% level in any model. The P-value of the gap $\beta_3 - \beta_2$ obtained by RE of the IV method is 10.7%; that is, the difference is almost significant at the 10% significance level. To further interpret the result, we can state that while a 1.0 [cent/litre] increase in the pre-tax price leads to a 0.948 [litre/person] reduction in gasoline demand, a 1.0 [cent/litre] increase in the specific tax rate induces a 1.201 [litre/person] saving, so the effect is greater by 27%.

Table 12.3 Estimated results for the linear specification eq. (8)

	Normal method		IV method	
	Fixed	Random	Fixed	Random
gdppc	4.2307***	4.4243***	0.6641	0.7028
(β_1)	(0.000)	(0.000)	(0.134)	(0.114)
pr + tr	−65.9300***	−66.0234***	−94.8156***	−94.7810***
(β_2)	(0.000)	(0.000)	(0.000)	(0.000)
tr	−10.5887	−16.4034	−22.5044	−25.3594
($\beta_3-\beta_2$)	(0.519)	(−0.325)	(0.151)	(0.107)

	Normal method		IV method	
	Fixed	Random	Fixed	Random
_cons	415.5407***	393.6381***	532.1231***	512.2656***
	(0.000)	(0.000)	(0.000)	(0.000)
R^2 within	0.2360	0.2359	0.3549	0.3559
R^2 between	0.6361	0.6377	0.6294	0.6305
R^2 overall	0.5177	0.5212	0.5283	0.5306
Number of ...	obs. 746, groups 29		obs. 571, groups 24	
IV			instrumented: trpr instruments: gdppc tr pcor	
Hausman Stat.	−20.71		-29.95	

Note: P-value of t-test is in parentheses; level of significance, *** 1%, ** 5%, * 10%.
Breusch-Pagan test revealed significance at 1% level, implying that the random effect panel is better than pooled OLS.

5. CONCLUSION

The effect of the tax rate on energy demand is stronger than that of a volatile market price. This is a fact that has seldom been proven empirically, except two studies. Our study clarified this fact by using the Panel Data method with data on gasoline demand, prices and taxes from 29 (or 24 by the Instrumental Variable method) countries.

While with a log-linear specification we obtained clear results in line with existing studies, with a simple linear specification, the difference between the price and tax effects was not necessarily significant, although the sign of the value thus obtained is consistently negative (this implies that the tax effect is stronger than the price effect). By estimation, we also dealt with the problem of endogeneity, which is more clearly shown in a log-linear specification.

The results of the log-linear specification significantly show that the elasticity of the tax-factor is 16.0% greater than that of the pre-tax price. Moreover, according to the results of linear specification, we can point out a similar fact if we are allowed to use the result of the RE model of the Instrumental Variable method for further interpretation. That is, we can state that while a 1.0 [cent/litre] increase in the pre-tax price leads to a 0.948 [litre/person] reduction in gasoline demand, a 1.0 [cent/litre] increase in the specific tax rate induces a 1.201 [litre/person] saving, so the effect was 1.27 times greater.

Several problems remain to be solved in the future. Firstly, we cannot distinguish between short-term and long-term tax effects. Secondly, we exclusively dealt with gasoline demand and price, while we did not consider other substitutes such as diesel oil. To consider this we may have to introduce a simultaneous equation system. Finally, we need to study whether consistency is also maintained with electricity, gas or other energy sources.

APPENDIX: DATA PROCESSING

We used the databases *Energy Prices & Taxes – Quarterly Statistics* and *Energy Balances of OECD Countries* by the IEA for this study. By dividing prices and tax rates expressed in local currencies by the purchasing power parity (ppp) dollar exchange rate, we obtained a series of prices and tax rates from 1978 to 2007 for 29 OECD countries. Panels of energy consumption and several macroeconomic indicators from 1960 to 2006 are also available. However, some values are missing for some years and for some countries; therefore, we only have 746 observations instead of 841 (29 years times 29 countries).

Price data are deflated by the consumer price index of the US.

In particular, there are many missing gasoline price values. The above-mentioned database, *Energy Prices & Taxes*, includes five price panels for gasoline, namely 'Premium Leaded Gasoline', 'Regular Leaded Gasoline', 'Premium Unleaded 95 RON', 'Premium Unleaded 98 RON' and 'Regular Unleaded Gasoline'. The observations for 'Premium Leaded Gasoline' are most numerous. However, there are more missing values in more recent years for many countries, as the usage of leaded gasoline gradually declined in many countries. Therefore, we used the second most popular category, 'Premium Unleaded 95 RON' for data after 1996 and substituted values of the original 'Premium Leaded Gasoline' for other remaining missing values. There have been price differences between the two categories of gasoline by a certain percentage. As leaded gasoline has been more expensive than unleaded on many occasions, the (newer) prices of 'Premium Unleaded Gasoline' have been multiplied by the ratio of the two gasoline prices in 1996 (or 1997).

For some countries, since many values remain missing even with the above-mentioned method, we used the original available data for 'Regular Unleaded Gasoline' expressed in ppp dollar. Specifically, the observations of this series are sufficient for these countries (Australia, Canada, Japan, Korea, Mexico, New Zealand and the US).

For tax rates, we primarily applied the 'excise tax' rate for 'Premium Leaded Gasoline', but substituted the tax rates for 'Premium Unleaded Gasoline 95 RON' or 'Regular Unleaded Gasoline' (depending on availability) for missing values. The discriminating tariff against leaded gasoline is applied in some countries. Therefore, to maintain consistency, we added the difference in tax rates between leaded and unleaded gasoline in the last available year, to the tax rate for 'Premium Leaded Gasoline'.

NOTES

[1.] In particular, $\ln(p + t)$ cannot be naively transformed into $\ln p + \ln t$, which indeed is the same as $\ln(pt)$.
[2.] We cannot use Iceland because it lacks tax-rate and price data.
[3.] Hausman statistics can have negative value. See Schreiber (2008).
[4.] By Panel Data analysis, it is generally assumed that coefficients are common between countries, although each country has a specific constant term.

REFERENCES

Bardazzi, R., F. Oropallo and M. G. Pazienza (2009) 'Industrial CO_2 emissions in Italy: A Microsimulation Analysis of Environmental Taxes on Firm's Energy Demand', in Cottrell J., J. E. Milne, H. Ashiabor, *et al.* (eds) *Critical Issues in Environmental Taxation VI*, New York, US: Oxford University Press, 465–91.

Ghalwash, T. (2007) 'Energy Taxes as a Signaling Device: An Empirical Analysis of Consumer Preferences', *Energy Policy*, **35** (1), 29–38.

Schreiber, S. (2008) 'The Hausman Test Statistics can be Negative even Asymptotically', *Journal of Economics and Statistics* (*Jahrbuecher fuer Nationaloekonomie und Statistik*), **228** (4), 394–405.

Wooldridge, J. M. (2006) *Introductory Econometrics*, Mason, US: South-Western Cengage Learning.

13. The CDM and the built environment

Javier de Cendra de Larragán

1. THE IMPORTANCE OF TACKLING ENERGY CONSUMPTION IN THE BUILDING SECTOR

Buildings worldwide account for 40 per cent of global energy consumption and resulting emissions, significantly more than the transport sector. At the same time, energy consumption and consequent emission are set to increase rapidly and dramatically, mainly in developing countries, due to population growth and increasing energy usage per person as a result of rising standards of living. Just to get an idea about the scale of the challenge, a few figures can be provided, as follows.

China is expected to add twice the amount of current US office space between 2000 and 2020. Also in China, 73 per cent of the population will live in cities, compared with 45 per cent now. Both China and India are developing new ways to manage the huge demand for new residential buildings. In China, 1 km² land parcels are being provided by cities with arterial streets in place. Developers build everything needed inside each one of those parcels, each holding 2000 to 10 000 housing units. Between 10 and 15 of these superblocks were being completed every day in 2008, adding 10 to 12 million new housing units per year. In India, new combined housing and office developments are being built in large land parcels on the outskirts of major cities. Some 400 township projects, with populations of 0.5 million each, are predicted over the next five years in 30 to 35 cities. Most of these buildings will still be standing by the end of the 21st century. Both in India and China, total energy consumption is expected to increase much more rapidly than energy efficiency. For instance, purchases of more air conditioners will result in increased total energy consumption, more than tripling by 2020 as compared with 2000 levels, even if a 40 per cent increase in efficiency is accounted for (Zhou *et al.*, 2007).

In order to avert the worst impacts of climate change, the International Energy Agency (IEA) has called for a reduction of 77 per cent in carbon emissions below business as usual (BAU) in all sectors of the economy by 2050 (IEA, 2006). The IEA also demonstrates that without substantial emission reductions in developing countries, it is impossible to achieve the 450 ppm CO_2-eqivalent scenario (IEA, 2008, p.478). According to the IEA, the building sector could contribute at global level 18.2 Gigatonnes of the total 48 Gigatonnes reduction required, and, as just seen, the opportunities in the developing world are massive, and in theory at a lower cost than in other sectors. At the same time, and as will become clear throughout this chapter, huge and very complex challenges will need to be overcome if those opportunities are to be realized, both in developed and developing countries.

Based on the very widespread assumption that it makes economic sense to invest in energy efficiency measures in the building sector, and given that the growth in emissions from that sector will mostly come from developing countries – which do not have binding mitigation targets under the Kyoto Protocol – many hopes have been placed in the potential of the Clean Development Mechanism (CDM) to facilitate that task.

However, 10 years after the start of the CDM, energy efficiency projects within the building sector are very limited in number and in scope,[1] and some effort has gone into investigating the underlying reasons as well as to examine whether amending the CDM in certain ways could help overcome this failure.

Against this background, this chapter seeks to explore whether and how the CDM is able to promote energy savings in the built environment. In order to do so, the chapter is structured as follows: Section 2 will briefly examine traditional explanations provided in the literature to explain the challenge of increasing energy efficiency in the built environment; Section 3 will assess the nature of the built environment with a view to identifying the existing barriers to achieve energy efficiency therein; Section 4 will assess the potential contribution and limits of the CDM to overcome those barriers; Section 5 will consider the contribution of a reformed CDM within a more comprehensive policy and regulatory approach to energy efficiency in the built environment; and Sector 6 will conclude with a number of recommendations to policy makers.

2. THEORETICAL APPROACHES TO EXPLAIN THE CHALLENGE OF REDUCING ENERGY CONSUMPTION IN THE BUILDING SECTOR

Traditionally, many economists argue that since investments to improve energy efficiency in buildings have negative costs, rational actors should make them, even in the absence of a carbon price. If anything, pricing carbon would increase the incentives to do so. So, if they do not make them, is because investors are factoring in the costs of overcoming all the barriers within the sector that raise transaction costs enormously (so-called hidden costs). Other economists have studied in depth the decision-making processes of economic actors, and have found that these are actually often very 'irrational', so that they fail to make investments even if they are really beneficial (Sorrell, 2009; Wilson and Dowlatabadi, 2007).

It must, however, be noted that all this work has remained until recently at a rather theoretical level. Comprehensive studies of the nature, structure and functioning of the building sector that would make 'hidden' costs visible to the analyst have been conspicuous by their absence, although they could throw light on the range and extent of market failures in the built environment.

3. The Complexity of the Building Sector and Consequences for Reducing Energy Consumption Therein: Some Empirical Findings

One important study carried out for the United Nations Environmental Programme (UNEP) and published in 2008 sought to understand in depth the barriers present in the building sector that prevent more energy savings to be achieved (Cheng *et al.*, 2008). The ultimate goal was to identify whether the CDM could be useful to overcome (some of) them. The barriers identified include (1) the huge amount of existing or planned buildings and the small volume of energy savings that can be achieved from each one of them, which makes achieving economies of scale very difficult; (2) the fragmentation and complexity of the construction sector, where there is poor integration among different key stakeholders such as developers, capital providers, engineers, contractors, agents, owners, users and local government; (3) the conservative bias of the construction sector, which is meant to shield investors from the high risks that inflict investments, particularly in fast growing developing countries; (4) split incentives between investors in energy efficiency projects an those that would actually benefit from the investments; (5) lack of information, asymmetrical information, and misinformation, which leads many within the sector to believe that energy-saving measures increase construction costs much more than they really do (apparently, information about energy efficiency options is often incomplete, unavailable, expensive to obtain and verify, which leads to poor in-depth knowledge about possibilities and costs); (6) general lack of expertise, management tools and indicators for energy management in buildings (this affects both the construction and operational phases); and (7) high transaction costs of investment projects in energy efficiency in buildings.

Just before this study was published, the World Business Council on Sustainable Development (WBCSD) issued a report entitled 'Energy Efficiency in Buildings: Business Realities and Opportunities' (WBCSD, 2007a). This was the first part of a two-part, very ambitious study, carried over a period of four years and with a budget of $15 million, to examine in depth the barriers to promoting energy savings in the building sector and to generate solutions to overcome them. The study included an in-depth collection of data and extensive modeling across the building sector in the six countries/regions with the largest building sector worldwide (the US, Japan, China, Brazil, India, and the EU). Its initial findings were taken into account in the abovementioned (Cheng *et al.*, 2008) report, so there are no differences therein. Nevertheless, the approach between the two studies was slightly different, because in the WBCSD study the focus was not on how to reform the CDM to increase its capacity to promote energy efficiency in buildings, but rather to explore barriers to increase energy savings in the built environment worldwide. The report was based on a combination of existing research, stakeholder dialogues, and a market research study to measure the stakeholder perceptions of sustainable buildings around the world. It was meant to establish a baseline of existing facts and trends that would be used in the future to develop scenario planning and modeling approaches to assess the needed actions to change energy consumption in buildings. The main facts found in the study which prevent more energy efficiency in the sector come in two basic types: (1) the enormous complexity of the sector, where the value chain is very fragmented and not integrated, with the result that each actor focuses on promoting its own interests, often without regard for the impacts on energy consumption. This applies to all actors from local authorities to capital providers, developers and users; (2) barriers within the industry, including underestimation of the contribution of buildings to climate change, overestimation of the costs of saving energy, lack of know-how and experience among professionals, lack of conviction and personal commitment on energy, and lack of leadership.

The main conclusion is that appropriate policies and regulations are essential to achieve market changes. Policies need to do three basic things: (1) establish a carbon price through tax, trading or regulation; (2) promote new technologies in the sector; (3) remove barriers to behavioral changes, for instance, through information and standard setting. More in particular, an effective policy framework needs to include issues such a holistic master urban planning, more effective building codes to enforce minimum required technical standards, information and communication to overcome the lack of know-how and to highlight the energy performance of individual buildings, incentives, including tax incentives, to encourage energy efficiency in building equipment materials and occupant consumption, accurate energy pricing, improve the degree of enforcement and introduce and enforce measurement and verification policies.

In 2009, the WBCSD issued the second part of its report, effectively a year after the UNEP report was published (WBCSD, 2009b). The WBCSD report is very clear about the prevailing sense of 'false' optimism within policy makers and professionals regarding the apparently huge opportunities that exist, at negative or very low costs, to dramatically improve energy efficiency in buildings. Then the report goes on to show how very profound changes need to be adopted in order to tackle the extreme complexity of promoting emission reductions in the building sector, and makes recommendations to overcome them.

General recommendations, which apply to *both developed and developing countries*, include:

1. Strengthen energy efficiency requirements and labeling in codes for increased transparency.
2. Enhance effective enforcement of building codes. Low level of compliance with energy efficiency requirements is a serious problem both in developing and developed countries.
3. Developing energy measurement (for each household) and labeling mechanisms requiring non-residential building owners to display energy performance levels.
4. Put in place or restructure tax incentives and subsidies to enable energy-efficiency investments with longer payback periods.
5. Encourage integrated design approaches and innovations. Property developers need to be encouraged to restructure business and contractual terms to involve designers, contractors, utilities and end-users early and as part of an integrated team.
6. Use public procurement laws to provide incentives for energy efficiency improvements in a holistic way.
7. Develop and use advanced technology to enable energy-saving behaviors.
8. Develop workforce capacity for energy saving.
9. Mobilize citizens to change energy-related behaviors, so that not only energy efficiency is promoted, but also absolute energy consumption is cut. Currently, substantial gains in energy efficiency are lost to increases in energy consumption per capita, in what is known as the rebound effect (Sorrell, 2009, p. 352 *et seq.*).

4. WHY THE CDM CANNOT BY ITSELF GENERATE ENERGY SAVINGS IN THE BUILT ENVIRONMENT

The UNEP report examined not only barriers to increased energy efficiency in the building sector, but also the reasons that could explain why energy efficiency in building projects (EEB projects) remain virtually absent from the CDM portfolio. The report finds that limitations in the CDM's modalities, procedures and methodologies are contributing to the CDM's under-utilization. This would include:

- Legal uncertainty regarding the definition of additionality employed by the CDM Executive Board (CDM EB), which creates uncertainty about the acceptability by the CDM EB of proposed projects.
- Difficulties related to technology-based methodologies – which are reductionist and micro-managing in their use of technology-by-technology and measure-by-measure project controls, and therefore generate very large transaction costs.
- Difficulties in establishing baselines for new buildings. Establishing baselines requires that data be collected from comparable buildings, which are by definition either non-existing or very scarce.
- Difficulties in establishing baselines for existing buildings, because buildings tend to be different from each other and because monitoring data for the target buildings may be lacking.
- Difficulties in evaluating the thermal performance of buildings. The problem is that many measures that improve the thermal performance of buildings are structural (orientation, passive cooling, building shape, shade), which are difficult to classify as specific technologies for validation, monitoring and verification purposes, and therefore do not fit existing CDM methodologies.
- The combination of different methodologies is not allowed for programmatic CDM, thus a variety of energy-saving approaches cannot be deployed even if this would be the most effective and cost-effective approach.
- Restrictions on recognizing soft measures to save energy, such as changing consumer behaviors and raising energy awareness, due to the fact that the CDM is technology based and thus cannot accept non-technical solutions. These measures would include using good standard operation procedures (SOPs), proper commissioning, good maintenance, optimizing operational conditions, record-keeping, linking energy savings to professional evaluation and advancement,

using energy-consuming devices on an as-needed basis, providing proper consumption information feedback and learning how to change consumption patterns and save energy.

The approach of UNEP's report needs to be contrasted with the findings of the WBCSD report. Doing so helps placing the role of the CDM in promoting energy efficiency in the built environment in perspective.

Indeed, both the UNEP and the WBCSD reports largely agree on the complexities to be found in the building sector, although the latter is far more detailed in this regard. However, the WBCSD puts the finger in the wound when it states that, according to extensive modeling, all the emissions reduction potential in the building sector should indeed be taking place at current carbon prices. In other words, the main barrier is not one of prices, but rather the combination of lack of effective regulations and poor enforcement in the sector, the complexity of the construction chain, lack of in-depth knowledge about the technologies and strategies to achieve further mitigation, lack of awareness about the contribution of buildings to the greenhouse effect and, last but not least, lack of expertise in modeling the energy consumption of buildings, including the energy relevant behaviors of indwellers. The report finds that increasing substantially carbon prices will only have a marginal effect on energy efficiency gains, given the number and depth of market failures. These market failures can only be overcome through a huge regulatory effort.

So can the CDM deliver in the absence of such effort? In essence, all the CDM does is to increase the incentives of potential investors by setting up a carbon price, which is just one of the dimensions of the regulatory effort required. So the CDM is simply not enough, unless the carbon price would be so high that it would cancel out the transaction costs generated by all the existing market failures, which is in any case an unrealistic scenario. And even then, as the WBCSD report acknowledges, one single actor cannot change the structure and organization of the entire sector. Although the report for UNEP acknowledges that the CDM needs to be seen within the wider picture in the post-2012 context, it does not seem to stress that point to a sufficient extent.

5. THE ROLE OF A REFORMED CDM WITHIN A COMPREHENSIVE REGULATORY MIX

5.1. Marginal Reforms to the CDM

In the light of the foregoing, it would seem that placing too much emphasis on reforming the CDM without giving sufficient attention to the broader picture is not a satisfactory approach. This conclusion can be reinforced if we note that the UNEP report explicitly refers to the need to develop national regulations and standards in the post-2012 World.

A better approach would be to ask whether the CDM could only be effective once a comprehensive regulatory framework is in place, or whether a properly reformed CDM could actually assist in overcoming existing barriers by helping to develop a comprehensive regime. The UNEP report seems to share the latter view. To that end, the report suggests a number of amendments to the CDM, some of which could be implemented in the short term, while the implementation of others would require making amendments to the Kyoto Protocol.

Those changes that do not require an amendment to the Kyoto Protocol include:

1. moving away from technology-based methodologies for small end-use energy efficiency projects and fully embracing the programmatic CDM
2. allowing performance-based methodologies
3. allowing the development of standardized baselines
4. improving additionality tools and requirements
5. strengthening the role of Designated National Authorities.

These changes would improve economies of scale, reduce transaction costs and increase certainty for developers that the CDM EB would approve the projects. Thus they would make CDM projects in the built environment a more attractive proposition, by removing some of the current barriers. Their most important contribution to overcoming existing barriers in the sector would seem to be their contribution to generate and spread knowledge and best practices about how to improve energy savings in the built environment of developing countries, so that other investors can replicate the measures. This could create more awareness within those countries regarding the benefits of undertaking a comprehensive approach to overcome existing barriers, thereby reaping the enormous (and currently hidden) efficiencies.

5.2. Expanding the CDM

In addition, there are proposals to expand the scale of the CDM. These include sectoral CDM, sectoral crediting, and sectoral trading, with either no-lose or binding targets. The objective of proposals for scaling-up the CDM is quadruple: (1) to increase global mitigation; (2) to increase cost-effectiveness; (3) to address carbon leakage; and (4) to promote sustainable development in developing countries.

The concept of sectoral CDM has been used in the literature to refer to different things, including policy-based projects, bundling of private projects and sectoral baselines or sectoral crediting. The main difference between the CDM and sectoral CDM is that the CDM typically applies to a single project, which is usually related to a single installation, while sectoral CDM can apply to several projects, sectors or policies. The main difference between sectoral CDM and sectoral crediting is that whereas the former would fall under the current legal and institutional framework regulating the CDM, the sectoral baselines would have to be negotiated by the Conference of the Parties (COP) level. Among options for sectoral crediting, an important alternative is the so-called 'sector no-lose targets' (SNLT). SNLT are non-binding crediting mechanisms that can be applied at sectoral level, at least for some sectors and countries, and which can encourage sector-wide emission reductions. Reductions below the baseline would generate credits that can be issued to the government, but no penalties would occur if the target is not met for the whole sector. SNLT are seen as a more realistic step than imposing upon developing countries absolute or even intensity-based, economy-wide targets. The latter would require allowing developing countries a large amount of hot air, which may not be palatable for developed countries, and which may moreover be counterproductive when developing countries lack the necessary institutional, legal and technical capacity to make a wise use of the revenues. SNLT represent therefore an intermediate step between the CDM and mandatory cap-and-trade schemes.

According to some, reductions below BAU and the baseline should not generate credits. This view sees SNLTs as going beyond offsetting. Whereas under the CDM additionality is assessed by the CDM EB by applying certain methodologies and includes investment additionality, additionality in SNLTs would depend on the baseline negotiated by a government, and would only have an environmental dimension. This solution, however, does not address the charge that SNLT in fact impose absolute emission-reduction targets upon developing countries, because they only generate credits for their action if they reduce below the baseline, not just below BAU. The EU and others have proposed that some reductions below BAU and above the baseline could be supported with finance from the international community. This position of the EU could constitute a middle ground to deal with the problem of additionality as mentioned above.

SNLTs, however, impose additional challenges of their own, as follows.

First of all, they are unlikely to be feasible for many key sectors, and moreover some developing countries may lack the capacity to develop SNLTs. SNLT requires setting accurate baselines that can be monitorable, reportable and verifiable (MRV) and eventually enforced, so they might be feasible for industrial sectors with small numbers of large sources, such as electricity generation, cement, aluminium or steel production and upstream emissions of oil and gas production. Countries that want to develop SNLTs need to be able to gather all the necessary data and develop the requisite MRV national system capacities. Hence, SNLTs are particularly appropriate for rapidly industrializing (or developing) countries where there is a need for significant investment, and where investments are otherwise likely to follow high-carbon patterns.

Second, there is the choice of baselines itself, which is likely to be a political exercise: too high and a huge oversupply of credits might be generated. Possibly, the choice of baseline should not be left to individual countries alone, but negotiations should be aided by an independent and objective technical advisory group set up by the COP. This group could also advise on absolute mitigation targets for Annex I countries. It would be the competence of the MOP to elaborate modalities and procedures for the review and approval of proposals and for MRV of emissions and of accounting of units.

Third, in understanding the scale of the task, it is necessary to consider that, while Annex I countries have had more than 10 years to prepare for emissions trading, many non-Annex I countries have only developed limited experience with CDM projects. Moreover, the EU has not yet been able to develop and agree upon EU-wide benchmarks in order to determine allocation of allowances to the installations covered by the EU ETS, despite the years and resources spent in that effort.

Hence, a substantial amount of capacity building will be required to gather data, choose relevant sectors, set up MRV requirements, develop a credible enforcement mechanism, set baselines, and so on ...

Last but not least, SNLTs have the potential – even with adequate baselines – to generate a very large amount of credits, and therefore a potential mismatch between supply and demand may arise if Annex I countries do not adopt sufficiently stringent mitigation targets. The EU alone would not be able to absorb all the increase in demand, hence stringent mitigation targets in other developed countries are also required. This last observation suggests that including the built environment in the carbon markets requires that large developing countries accept absolute emissions caps, at least at sectoral level, but preferably at national level. Otherwise the market would be flooded with cheap credits and would lose its environmental rationale.

5.3. Can an Expanded CDM Work for the Built Environment?

It has been said that, since sectors such as buildings and transport have large numbers of small sources, a SNLT would be very complex and perhaps unfeasible – although some sub-sectors could be defined, including perhaps regions of sub sectors that are less than national in scale (Streck and Winkler, 2008). However, provided that some changes are made, it could work for buildings. UNEP has suggested that allowing performance based methodologies, which focus, for instance, on measuring energy performance per square meter rather than on measuring improvements technology by technology and measure by measure, offers a good alternative means of accommodating energy end-use projects, which are characteristically small in scale and large in quantity. In order to use performance-based methodologies, it would be necessary to establish performance-based common baselines, which could then be coupled with energy saving targets for 'end-users'. As long as they overcomply with their targets, they could generate carbon credits that could sell in the international market. These two issues will be examined in turn.

To start with, developing common baselines faces relevant technical challenges that in turn would have legal consequences.

As the report for UNEP states, currently there are computer simulation programs and other tools that allow constructing baselines using data on building types and materials, orientation, cooling and heating-degree days, etc., in different countries and climate zones. The key problem, particularly for developing countries, is the lack of available data. Estimations and extrapolations based on developed country data may be used for certain building types (e.g., commercial air-conditioned buildings), as long as CDM authorities could accept the methods used and results. Large-scale surveys would be needed in order to obtain data sufficiently realistic to establish baselines that meet the requirement imposed by Article 12 of the Kyoto Protocol regarding the need that emission reductions be real and measurable. Because of the diverse construction and energy use patterns in residential buildings, common baselines would be especially helpful to residential building projects seeking to participate in the CDM.

Of course, there is a risk that common baselines would not be representative of all building types and moreover could not take into consideration local conditions and future variations. This may render them unacceptable in the context of the CDM. Refining and adjusting baselines according to data availability, technology developments and changes in CDM crediting approaches, however, could solve technical concerns. Given the difficulties inherent in setting up projects in the building sector, the role of the CDM here seems very limited unless CDM regulators and authorized entities provide assistance for baseline development. The CDM has used a similar approach to solve problems in areas where project proponents have had difficulty obtaining data. For example, to facilitate the calculation of emissions from electricity use, commonly accepted emission factors published by local utility authorities are used to calculate emission reduction baselines. In many cases, the emission factors are 'average' values that do not strictly reflect the actual emissions of the electricity producers from which the project draws power.

Second, and in relation with targets, it could be possible to learn from discussions taking place within the EU in relation to the effectiveness of setting up energy saving targets for 'end-users' in some sectors (Wesselink and Harmsen, 2010) . If those users manage to overcomply with those targets, then they would be able to generate carbon credits that could sell in the international market. In this way, SNLT would perform similar role than the legally binding energy saving targets at 'end-user' level currently being proposed within the EU.

6. SOME RECOMMENDATIONS TO REDUCE ENERGY CONSUMPTION IN THE BUILT ENVIRONMENT OF DEVELOPING COUNTRIES

According to the report for UNEP, it is essential to ensure that the experience gained by developed countries in promoting energy savings in the built environment is transferred as soon as possible to developing countries, and ideally instantaneously. In particular, sharing experiences learned in countries that have already established policies and regulations would certainly help in the development of financially supported national policies.

This recommendation has to be put in the context of the current international climate change negotiations. One of the challenges there is to induce developing countries to engage into mitigation actions while not putting their legitimate development goals at risk. Currently, developing countries are only involved in mitigation through the CDM, but other forms of involvement are conceivable, such as expanding the CDM, as already seen. The Bali Action Plan[2] called for 'nationally appropriate mitigation actions (NAMAs) by developing country Parties in the context of sustainable development, supported and enabled by technology, financing and capacity-building in a measurable, reportable and verifiable manner'. This text involves some of the key issues within the current negotiations. There are still many ill-defined issues in relation to NAMAs, including their legal nature, the difference between actions and commitments (of developed countries), their scope and the link between NAMAs and the support received from developed countries. The latter is the crucial issue in the context of this chapter, because it has been shown that promoting massively energy efficiency in the building sector is not something within reach of almost any country in the world, let alone developing countries. The Copenhagen Accord adopted in 2009 does also refer to NAMAs, by encouraging developing countries to list their NAMAs . The Accord mentions a registry listing NAMAs seeking international support along with a basic framework for doing so. Many developing countries have communicated their NAMAs to the United Nations Framework Convention on Climate Change (UNFCCC) Secretariat.[3] NAMAs can include policies in the building sector, which could be linked to international support. Moreover, the elaboration of NAMAs themselves can profit from assistance from developed countries, which seems particularly important in the context of the built environment. In the light of these observations, a number of suggestions on how to increase the role of buildings within NAMAs are hereby put forward:

- First, some developed countries are putting considerable effort both in getting and processing enough data on the energy behavior of buildings to feed models, and in developing further models for setting baselines and estimating the energy behavior of buildings, given that these models are still rather unsophisticated. Moreover, within the EU there is a push to roll out a smart grid that will include smart meters, which will provide accurate and real-time data on consumption of energy per dwelling and therefore will allow monitoring with a high degree of certainty the effectiveness of energy efficiency measures. These improvements, transferred to developing countries, would enable the extension of sectoral CDM to the built environment, hence increasing substantially economies of scale therein. It seems crucial then that, when developed countries commission research in these areas, they consider from the outset the potential relevance of their work for developing countries. At the moment research commissioned by the UK and by the European Commission does not seem to be paying sufficient attention to this issue.[4]
- Second, developed countries should make sure that any findings are then transferred to developing countries, for instance, by encouraging the early participation of universities from developing countries in

research consortia. Indeed, research institutions in developed countries with extensive experience in gathering data and in developing methodologies could cooperate further with research institutions in developing countries.

- Third, capacity building in designing an effective regulatory framework is essential, so that developing countries can design building regulations that are adequate for their circumstances and to increase monitoring and enforcement capabilities. This may require assisting with the design of laws and/or changes to existing laws to facilitate the role of courts in enforcing the building regulations. Indeed, without effective enforcement the sector cannot hope to make a leap forward in energy efficiency.
- Fourth, effective regulatory frameworks need to be implemented, and this does not only depend on political will and sufficient resources, but also on the workings of, mainly, domestic administrative and tort law. Much more work is needed to explore how existing laws could be amended to improve enforcement. The role of tort law in particular to provide incentives to comply with building regulations could be key.
- Fifth, contractual solutions need to be exported fast within the industry. Given its atomized approach, this may also require assistance from governments through capacity building programmes. A fundamental transformation of the ways in which the sector operates at global level is needed fast. Also in this there is a need to transfer experiences from leading countries.

It must be noted that developed countries have a real interest in ensuring that all their knowledge is swiftly transferred to developing countries for a very simple and important reason: after all, their building sector extension is negligible compared to the phenomenal growth in countries such as Brazil, India, China, and others. All their efforts would be ineffective in reducing aggregate emissions, and therefore climate change, if they are not made useful for developing countries. And since many of the barriers to promote more energy efficiency in the sector are of the sort that needs to be tackled by public bodies, it makes sense to deploy public aid to transfer new knowledge to developing countries. In other words, issues of intellectual property may not arise here.

To conclude, it would be unwise to put too much hopes on the contribution of the CDM to the decarbonization of the built environment in developing countries. Instead, there is a need to tackle head on – through a comprehensive regulatory strategy – all the barriers that afflict the sector by using the experience gained in leading developed countries. Only then will the carbon market work effectively. At the same time, it is necessary to be realistic; developed countries are turning their attention to energy savings in the built environment after having developed their climate policies for almost 20 years (since the adoption of the UNFCCC). Only after having harvested the 'low-hanging fruit' are they tackling much more challenging trophies. Should developing countries be expected to do otherwise, and could they? In the meantime, it is important to ensure that lessons learned in developed countries are swiftly transferred to developing countries.

NOTES

1. These projects can be seen at http://cdmpipeline.org/cdm-projects-type.htm. Accessed 5 December 2010. Projects include, e.g., a project that reduces energy consumption in a hotel in India (Project 0686) and several projects on energy-efficient lightning. Total certified emission reductions (CERs) coming from energy efficiency projects in the demand side represent around 0.3 per cent of all CERs issued so far, and they are expected to represent around 1 per cent of all CERs by 2012.
2. Decision 1/CP. 13 Bali Action Plan, FCCC/CP/2007/6/Add.1.
3. A list of NAMAs is available at http://:unfccc.int/home/items/5265.php. Last accessed 5 December 2010.
4. In the UK, research in the energy efficiency of buildings is commissioned mainly by the EPSRC, http://www.epsrc.ac.uk/Pages/default.aspx. The Commission's 7th Framework Programme can be consulted at http://cordis.europa.eu/fp7/home_en.html.

REFERENCES

Cheng, C., S. Pouffary, N. Svenningsen (2008), The Kyoto Protocol, the Clean Development Mechanism, and the Building and Construction Sector, Paris, UNEP.

International Energy Agency, World Energy Outlook 2006.

International Energy Agency, World Energy Outlook 2008.

Sorrell, S. (2009), 'Improving Energy Efficiency: Hidden Costs and Unintended Consequences', in Dieter Helm and Cameron Hepburn (eds), The Economics and Politics of Climate Change, Oxford, Oxford University Press, pp.340–61.

Strech, C., and H. Winkler (2008), The Role of Sector No-lose Targets in Scaling Up Finance for Climate Change Mitigation Activities in Developing Countries, London, Defra.

Wesselink, B. and R. Harmsen (2010), Energy Savings 2020: How to Triple the Impact of Energy Saving Policies in Europe, Brussels, European Climate Foundation.

Wilson C. and H. Dowlatabadi (2007), 'Models of Decision Making and Residential Energy use', Annual Review of Environment and Resources, 32, 169–203.

World Business Council for Sustainable Development (2007a), Energy Efficiency in Buildings: Business Realities and Opportunities, Geneva, Washington, Brussels.

World Business Council for Sustainable Development, (2007b), Energy Efficiency in Buildings: Transforming the Market, Geneva, Washington, Brussels.

Zhou, N., M. McNeil, D. Fridley, et al. (2007), Energy use in China: Sectoral Trends and Future Outlook, Lawrence Berkeley National Lab.

14. CGE analysis of border tax adjustments

Masato Yamazaki

INTRODUCTION

At the UN summit in 2009 the Japanese government pledged to reduce domestic CO_2 emissions by 25 per cent by 2020, relative to the 1990 levels. To achieve this target it plans to introduce a nationwide emissions trading scheme for CO_2 in the near future. However, implementing a unilateral or sub-global emissions trading scheme in Japan could cause a drastic increase in the production costs of Japan's carbon-intensive sectors, and on the domestic and foreign markets, these sectors could suffer from reduced competitiveness relative to companies in countries without comparable emissions reduction policies.

Therefore this is a debatable issue. First, it would be unfair that the competitiveness of carbon-intensive sectors of countries with less or no CO_2 emissions regulation would be increased by Japanese regulations on CO_2 emissions. This unfairness, in turn, could lead to strong domestic opposition that could undermine support for the implementation of emissions trading. Second, so-called 'carbon leakage' might occur. In other words an increase in the competitiveness of carbon-intensive sectors in countries with less or no regulation could lead to an increase in CO_2 emissions in those countries because of increased production.[1] To avoid the problems associated with unilateral or sub-global emissions trading, border tax adjustments (BTAs) to complement emissions trading schemes are currently being discussed in developed countries. BTAs impose taxes on imports on top of ordinary tariffs. In principle the tax rate is decided on the basis of the amount of CO_2 emitted by the production of imports in the exporting country. Thus BTAs for imports are sometimes called 'carbon tariffs'. In addition BTAs include previously paid emissions cost rebates on exports to countries with less or no regulation. In principle the rebate rate is decided on the basis of the amount of CO_2 emitted by the domestic production of exports.[2] In other words BTAs are government trade measures that could ensure fair competition between countries with regulation and countries with less or no regulation. The implementation of BTAs is also expected to mitigate carbon leakage.[3]

There are many issues that should be considered with regard to BTAs, such as the compatibility of BTAs with World Trade Organization (WTO) law and the effects of BTAs on carbon leakage. This chapter, however, focuses on the domestic economic impact of implementing BTAs to complement an emissions trading scheme in Japan. In particular the chapter examines the effects of BTAs on the production of domestic sectors, Japan's GDP and the price of emissions permits. In addition this chapter investigates the economic implications of BTAs by comparing their effects with the effects of sectoral exemptions. The policy of sectoral exemptions releases specific industries from the obligation to purchase CO_2 emissions permits.

In this chapter it is assumed that steel products imported into and exported from Japan are subject to BTAs. The iron and steel sectors are carbon-intensive. Moreover the Japanese iron and steel sectors face fierce international competition from corresponding sectors in countries with less or no regulation. The sectoral exemptions considered in this study are a policy that releases the iron and steel sectors from the obligation to purchase CO_2 emissions permits.

This investigation employs a computable general equilibrium (CGE) model that is a static, single-country, multisector model focusing on the Japanese economy. CGE models are widely used in economic simulations to assess the economic impact of climate change policies and trade policies. As will be explained later this chapter's CGE model treats the iron refining and steelmaking processes in detail.

COMPETITIVENESS CONCERNS OF JAPAN'S CARBON-INTENSIVE SECTORS

This chapter assumes that imported and exported steel products are subject to BTAs that complement an emissions trading scheme in Japan. The following are the reasons the study focuses on these products.

First, the iron and steel sectors are among the most energy-intensive sectors and are major contributors of CO_2 emissions in developed countries, including Japan. Therefore the implementation of an emissions trading scheme would have a large negative impact on these sectors; this impact must be examined thoroughly in climate policy analysis.

Second, Japanese iron and steel products face particularly fierce international competition, especially in foreign markets. Figure 14.1 displays the changes in the quantity of Japanese steel production as well as in the quantities of exports and imports of iron and steel products from 2000 to 2008. While imports of these products currently account for only a small share of the iron and steel markets in Japan, a large proportion of domestically produced iron and steel products – about one-third – is exported. Export markets are one of the most important sources of profit for Japan's iron and steel sectors. Figure 14.2 shows the percentage breakdown of countries that exported iron and steel products to Japan in 2008, while Figure 14.3 shows the countries that imported iron and steel products from Japan in the same year. The pie charts show that South Korea, China and other Asian countries are important trade partners of Japan for iron and steel products.

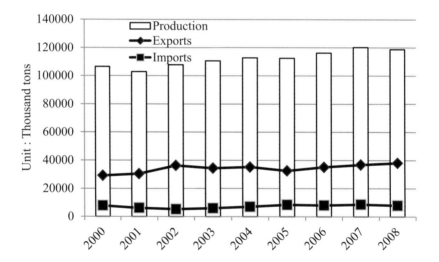

Source: The Japan Iron and Steel Federation (2009).

Figure 14.1 Changes in volume of steel production, exports and imports in Japan

Table 14.1 displays the top 10 ranking of the world's largest steel-producing companies for 2005 and for 2008. In 2005 there was only one Chinese company among the top 10. However, in 2008 four Chinese companies appeared in the top 10. Many Chinese steel companies are poised to overtake Japanese steel companies. In addition China is classified

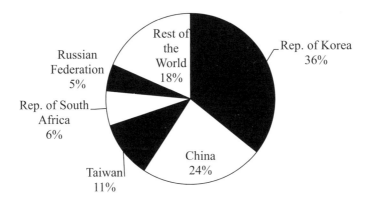

Source: UN Comtrade.

Figure 14.2 Share of countries exporting to Japan in 2008

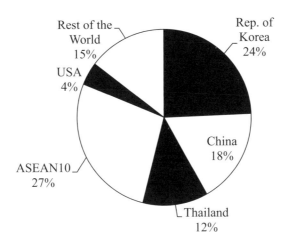

Source: UN Comtrade.

Figure 14.3 Share of countries importing from Japan in 2008

as a non-Annex I country in the United Nations Framework Convention on Climate Change (UNFCCC). Therefore it does not have any obligation to reduce its domestic CO_2 emissions under the Kyoto Protocol. Moreover it is highly unlikely that China and other Asian countries will take domestic action to reduce emissions at the same level as Japan after the Kyoto

Protocol. It follows that Japanese iron and steel companies will be threatened by competitive iron and steel companies in other Asian countries, especially China, in world markets for steel products.

Table 14.1 Top global steel-producing companies in 2005 and 2008

Rank	Firm	Share in 2005	Rank	Firm	Share in 2008
1	Mittal Steel (Nederland)	4.41%	1	Arcelor Mittal (Luxembourg)	7.79%
2	Arcelor (Luxembourg)	4.12%	2	Nippon Steel (Japan)	2.78%
3	Nippon Steel (Japan)	2.91%	3	Baosteel Group (China)	2.67%
4	POSCO (Rep. of Korea)	2.78%	4	POSCO (Rep. of Korea)	2.62%
5	JFE (Japan)	2.61%	5	JFE (Japan)	2.55%
6	Baosteel Group (China)	2.01%	6	Hebei Iron and Steel Group (China)	2.51%
7	US Steel (USA)	1.70%	7	Wuhan Steel Group (China)	2.09%
8	Nucor (USA)	1.63%	8	Tata Steel (India)	1.84%
9	Corus Group (UK)	1.61%	9	Jiangsu Shagang Group (China)	1.76%
10	Riva (Italy)	1.55%	10	US Steel (USA)	1.75%

Third, other carbon-intensive sectors in Japan are not exposed to comparably fierce international competition. Monjon and Quirion (2010) discuss whether major carbon-intensive products, such as aluminum, cement and iron and steel products, are candidates for subjects of border adjustments in the EU. They conclude that the answer depends on the EU's volume of imports and exports of these products. In Japan electrolytic refining of alumina disappeared due to the high price of electricity after the 1970s oil shocks. Consequently BTAs for aluminum products would not be needed at all under Japan's emissions trading scheme. In contrast the Japanese cement sector is not exposed to international competition, for almost all cement products consumed in Japan are domestically produced, and the volume of cement product exports is small. Thus BTAs would also not be needed for cement products. This is the third reason this chapter's study assumes that BTAs are set for imports and exports of steel products in the simulation analysis.

Another feature of this study is that it considers the economic implications of BTAs by comparing them to sectoral exemptions. Sectoral exemptions have been used to protect industries that are exposed to international competition from the burden of environmental taxation; for instance, they have been implemented in Norway and Sweden. In this study sectoral exemptions release the iron and steel sectors from the emissions trading scheme for CO_2 in Japan. In other words the exempted iron and steel sectors do not have to purchase emissions permits under the emissions trading scheme. In addition to BTAs, sectoral exemptions seem to be one of the most straightforward ways for the Japanese iron and steel industries to maintain international competitiveness.

General Description of CGE Models

This section begins with a general description of CGE models, which are economic simulation models that are widely used to assess the economic impacts of climate change policies and trade policies. CGE models are based on the general equilibrium theory of economics, and each such model consists of a system of nonlinear simultaneous equations that represent market interactions between economic agents such as households, industries (production sectors) and governments.

The interactions that are described by CGE models are depicted in Figure 14.4. The agents in CGE models behave as follows. Households supply labor and capital to their respective markets to earn income. Households also maximize their utility through consumption choices subject to their income constraints. Production sectors supply products to each market and demand intermediate products, labor and capital to use in production. Production sectors also minimize their unit production costs through their choices of inputs. Government consumption is financed by taxes levied on the production sectors and on households. A portion of the products supplied to the domestic market is invested for domestic fixed capital formation. The level of investment mainly depends on national savings levels.

CGE models also incorporate foreign trade. Products supplied to domestic markets include not only domestic products but also imported products.

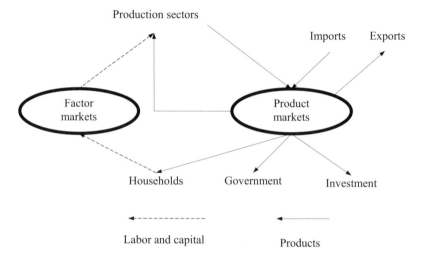

Figure 14.4 Flow of production factors and products in CGE models

Moreover a portion of domestic products is exported. However, a single-country model treats foreign countries as a single integrated virtual trade partner.

In general equilibrium theory, equilibrium is characterized by a set of prices and levels of production in each industry such that market demand equals market supply for all products. Price adjustments in each market ensure the equilibrium.

The parameters of the equations in a CGE model are estimated by using an input–output table that describes actual economic transactions. Parameter values are specified for the model to reproduce actual economic transaction data that are described in the input–output table as a benchmark (initial) equilibrium solution. This study examines how the benchmark values change in response to exogenous changes in the parameters of interest.[4]

Features of the model
This study's single-country CGE model focuses on the Japanese economy and consists of 38 production sectors, 53 products, three final demand sectors (consumers, the government and investors) and international trade (exports and imports). The parameters of the model's equations are estimated using the 2005 input–output table for Japan. One of the most important features of the model is that the iron and steel sectors are disaggregated in detail. The purpose of disaggregation is to capture differences in production methods and differences in types of steel products.

Figure 14.5 depicts the processes of iron and steel production as described by the model. There are two ways to produce steel in the model: the blast furnace–basic oxygen furnace (BF–BOF) method and the electric arc furnace (EAF) method. EAF steel products are produced primarily from steel scrap and are less carbon-intensive than BF–BOF steel products, whose main raw materials are iron ore and coke. EAF steel products are more scrap-intensive than BF–BOF steel products, although both use steel scrap. A BF–BOF steelmaker and an EAF steelmaker each produce five types of steel products. One category of steel products, 'steel plate and strip', includes steel products for car bodies and electrical appliances, which are produced primarily by the BF–BOF method. Another category, 'steel bar', includes steel products for reinforcing concrete, which are mainly produced using the EAF method. The classification of steel products is based primarily on the basic sector classification in the 2005 input–output table for Japan.

Structure of the model

The CGE model describes the production functions of the production sectors using a set of constant elasticity of substitution (CES) functions. In CGE models, CES functions, which are represented here by Equation (1), are widely used to model relationships between raw material inputs and their products. Suppose that there are N types of raw materials (1, ..., N). The independent variable X_i in Equation (1) represents quantities of the i-th raw material inputs. The subscript i denotes the type of raw material. The dependent variable Q represents the quantity of the product that is produced by the processing of N raw materials. θ_i is the share parameter of the i-th raw material inputs. σ in Equation (1) determines the degree of elasticity of substitution, which represents the percentage change in the relative input quantity demanded in response to a 1 per cent change in its relative price. The value of σ can range from 0 to ∞. If σ equals 0, Equation (1) indicates that there is no possibility of substitution among the raw materials. On the other hand if σ equals ∞, the equation indicates perfect substitution among the raw materials.[5]

$$Q = \gamma \left(\sum_i \theta_i X_i^{\frac{\sigma-1}{\sigma}} \right)^{\frac{\sigma}{\sigma-1}} \tag{1}$$

In CGE models, CES functions are usually nested to represent various input substitution possibilities. The nested production structure in this study's model is shown in Figure 14.6. Each industry in the model has input substitution possibilities among fuels (except for gasoline, jet fuel

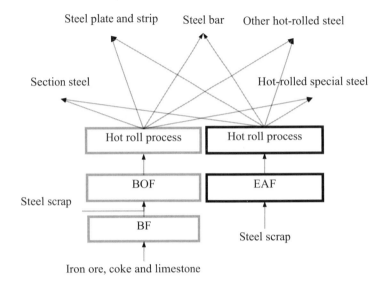

Figure 14.5 The production process of steel products in the model

and light oil, since no substitutes for these fuels exist in the industrial activities described by the model), between aggregate fuel and electricity and between aggregate energy and value added. The value added is the aggregate of labor inputs and capital inputs. Labor inputs and capital inputs are also substitutable for all industries. The model assumes that industries like the petroleum and chemical product industries that use fossil fuels or petroleum products as raw materials rather than as sources of energy have no input substitution possibilities for any of their intermediate products. Moreover in EAF crude steel production, no substitutes for electricity exist. The production structures of these industries are presented in Figure 14.7. In addition for all industries, there is no substitute for coke, although coke is used not only as a reducing agent in smelting iron ore in a BF process but also as a fuel. In the model, household consumption and investment in goods are also modeled using a set of CES functions. The nested structure of the CES functions for household consumption and investment is depicted in Figure 14.8. Government consumption is fixed at the initial level in the model. This study primarily uses the same values of σ that are used in the MIT Emissions Predictions and Policy Analysis model (Paltsev *et al.*, 2005). In each steel product market in the model, the market share of the EAF steel product or of the BF–BOF steel product changes in response to changes in relative price. The change in the share of each steel product is modeled using a technology-based logit approach represented by Equations (2) and (3). In Equations (2) and (3), the subscripts E and B represent the EAF and BF–BOF methods respectively, and j represents the type of steel product, such as steel bar, section steel, steel plate and strip, hot rolled special steel or other hot rolled steel. $S_{E,j}$ is the EAF product share in the market for steel product j, and $S_{B,j}$ is the BF–BOF product share in the market for steel product j. $\delta_{E,j}$ and $\delta_{B,j}$ are the base year quantities of the EAF and BF–BOF products respectively in the market for steel product j. $P_{E,j}$ and $P_{B,j}$ are the unit prices of the EAF and BF–BOF products respectively in the market for steel product j. λ_j is the substitution parameter represented by Equation (4), and its value represents a percentage change in the relative market share of the EAF and BF–BOF products in the market for steel product j in response to a 1 per cent change in their relative price. The model sets all λ_j to -1.5. This value is used in similar equations in the second-generation model to represent the substitution possibilities for EAF crude steel, BF–BOF crude steel and other methods in Germany (Schumacher and Ronald, 2007). The base case market shares of the EAF and BF–BOF products in each steel product market are established on the basis of information from the website of the Japan Metal Daily[6] and from annual steel production statistics (Tekko Shinbunsha, 2005).

$$S_{E,j} = \frac{\delta_{E,j} P_{E,j}^{\lambda_j}}{\delta_{E,j} P_{E,j}^{\lambda_j} + \delta_{B,j} P_{B,j}^{\lambda_j}} \tag{2}$$

$$S_{B,j} = \frac{\delta_{B,j} P_{B,j}^{\lambda_j}}{\delta_{E,j} P_{E,j}^{\lambda_j} + \delta_{B,j} P_{B,j}^{\lambda_j}} \tag{3}$$

$$\lambda_j = \frac{\partial \left(\dfrac{S_{E,j}}{S_{B,j}} \right) \left(\dfrac{P_{E,j}}{P_{B,j}} \right)}{\partial \left(\dfrac{P_{E,j}}{P_{B,j}} \right) \left(\dfrac{S_{E,j}}{S_{B,j}} \right)} \tag{4}$$

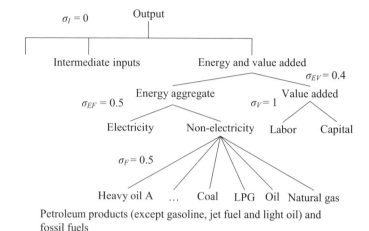

Figure 14.6 The production structure of the industries (excluding chemical, petroleum, coal, and EAF crude steel production)

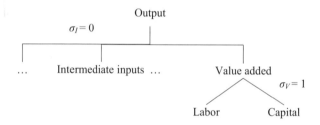

Figure 14.7 The production structures of chemical products, petroleum products, and EAF crude steel in the model

SIMULATION SCENARIOS

This study examines the economic implications of BTAs, particularly in comparison to the alternative competitiveness policy of sectoral exemptions. In the simulations, imported and exported steel products are subject to BTAs. Note that the analysis assumes that all imported steel products (section steel, steel plate and strip, steel bar, other hot-rolled steel, hot-rolled special steel, cold-finished and coated steel, cast and forged steel products and other iron or steel products) are produced by BF–BOF in exporting countries; thus they are all subject to BTAs. However, the

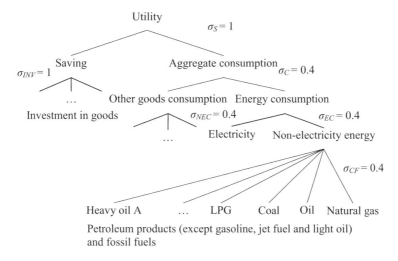

Figure 14.8 Structure of household consumption and investment

analysis assumes that emissions cost rebates for exports are applied only to steel plate and strip, cold-finished and coated steel, cast and forged steel products and other iron or steel products, which are mainly produced by BF–BOF in Japan. In the case of sectoral exemptions, only BF–BOF sectors, such as the production of pig iron using BF, production of steel using BOF and production using the hot-roll process for BOF steel, are exempt from the obligation to purchase CO_2 emissions permits under the emissions trading scheme.

   This study presents three policy scenarios. In the first scenario there is an emissions trading scheme without BTAs or sectoral exemptions. In other words in this policy scenario, the emissions trading scheme is implemented in the absence of any competitiveness policies. In the second policy scenario there is an emissions trading scheme with BTAs for steel products. BTAs impose carbon tariffs on imported BF–BOF steel products and grant emissions cost rebates for exported BF–BOF steel products. The rates of carbon tariffs on imported BF–BOF steel products and the rebates on exported BF–BOF steel products are calculated as the emissions permit price of one ton of CO_2 emissions in Japan multiplied by a CO_2 emissions coefficient for such products. The CO_2 emissions coefficient adopted by the model is the volume of CO_2 emissions per ton of crude steel production by the present average BF–BOF method (IEA, 2007). No sectors are exempt from the emissions trading scheme in this scenario. In the third policy scenario, there is an emissions trading scheme with sectoral exemptions for the Japanese BF–BOF sectors (that is, production of pig iron using BF, production of steel using BOF and production using the hot-roll process for BOF steel). In this scenario there are no BTAs for any imported or exported steel products. Note that the steel products manufactured by the exempted iron and steel sectors cannot be subject to BTAs because the simultaneous implementation of these policies for the same sector would be clearly contrary to WTO rules. In all policy scenarios the cap on CO_2 emissions is set for Japan to attain the target of a 25 per cent reduction in domestic CO_2 emissions relative to 1990 levels.

SIMULATION RESULTS

The simulation results for the three policy scenarios are as follows. Figure 14.9 shows the projected sectoral output impacts derived from each policy scenario. The black bars in Figure 14.9 indicate the case of emissions trading without BTAs or sectoral exemptions. Emissions trading without any competitiveness policies has a relatively large negative economic impact on the BF–BOF sectors; the production level of steel using BOF decreases by about 28 per cent. In contrast steel production using EAF increases by about 9 per cent, because steel product users substitute less carbon-intensive EAF steel products for BF–BOF steel products. Besides the BF–BOF sectors, thermal power generation, which uses fossil fuels as a source of energy, also drops dramatically because of the implementation of emissions trading.

The white bars indicate the case of emissions trading with BTAs. The BTAs mitigate the decrease in the BF–BOF steel production by one-third. In addition EAF steel production increases more than in the case of no competitiveness policies. However, the recoveries of steel production that result from the implementation of BTAs are accompanied by a decrease in the output of almost all other sectors. BTAs increase the price of imports and domestic products in the domestic market. This drives a price increase in raw materials. Second, the emissions permit market becomes tighter as a result of the steel production recoveries. Consequently other sectors have to bear additional production and emissions costs stemming from BTAs.

The projected sectoral output impacts of an emissions trading scheme with sectoral exemptions are represented by the bars with horizontal lines in Figure 14.9. The emissions trading scheme does not have a direct negative impact on the BF–BOF sectors. In contrast EAF steel production decreases by about 5 per cent because the EAF sectors do not have a competitive price advantage over the BF–BOF sectors under a policy of sectoral exemptions. Furthermore sectoral exemptions cause additional

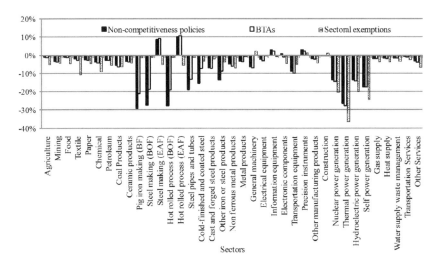

Figure 14.9 Changes in output levels compared with 2005 levels (%)

large decreases in the production levels of other sectors, except for steel-intensive industries, such as those that produce general machinery, electrical equipment and transportation equipment, and the construction industry. It should be noted that sectoral exemptions would exacerbate decreases in the production levels of products that do not contain steel, such as agricultural products, food products and textiles.

Figure 14.10 depicts changes in Japan's GDP under each policy scenario. In the simulation of the implementation of an emissions trading scheme without any competitiveness policies, Japan's GDP decreases by about 1.27 per cent compared with the case of no emissions reduction. The implementation of BTAs reduces Japan's GDP even more, by about 1.38 per cent compared with the case of no emissions reduction. Although BTAs would be partially effective in maintaining production levels of steel products, they would reduce national output levels. The implementation of sectoral exemptions depresses national economic activity even further; that is, GDP decreases by about 2.28 per cent compared with the case of no emissions reduction.

Figure 14.11 displays the projected price of an emissions permit. The recovery of the BF–BOF sectors with the implementation of BTAs would tighten the emissions permit market and cause an increase in emissions permit prices. Emissions permit prices would increase even more in the case of sectoral exemptions. The result of the implementation of sectoral exemptions would be that fewer industries covered by an emissions trading scheme would have to achieve the reduction target, which would remain at the same level as in the case of no sectoral exemptions. In other words under sectoral exemptions, reasonable options that exempted sectors may have for domestic emissions reduction must be overlooked. This causes the emissions permit price to rise.

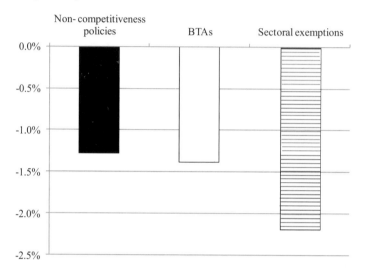

Figure 14.10 Changes in GDP compared with the case of no reduction (%)

Figure 14.12 shows changes in the projected market shares of imported steel products in Japan. The bars with vertical lines indicate the market share in the base case (the case of no reduction). As can be seen from the black bars, the implementation of emissions trading with no competitiveness policies would hamper the international competitiveness of Japan's steel products. The white bars show that the implementation of BTAs, in general, would reduce the market share of imported products to the market share of the base case. As can be seen from the bars with horizontal lines, the implementation of sectoral exemptions would also be effective in protecting the market share of domestically produced steel products. However, one should observe that there is no notable difference in the effectiveness of market protection between BTAs and sectoral exemptions.

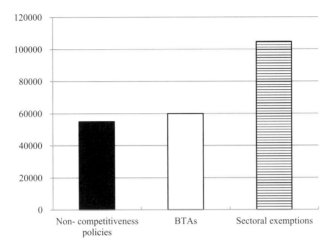

Figure 14.11 Price of an emission permit (JPY)

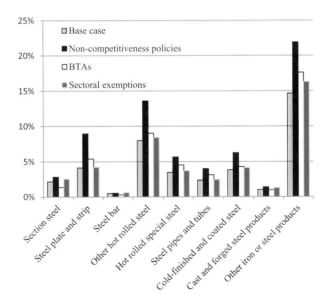

Figure 14.12 Changes in the market shares of imported steel products (%)

CONCLUSION

The international competitiveness concerns of the Japanese iron and steel sectors arise from Japan's pledge to reduce domestic CO_2 emissions by 25 per cent relative to the 1990 levels by 2020. This chapter examined the effects of two competitiveness policies that could serve as complements to the emissions trading scheme in Japan: BTAs for imported and exported BF–BOF steel products and sectoral exemptions for the BF–BOF sectors. The study used a static, single-country CGE model that focused on the Japanese economy.

The conclusions of the simulation study are as follows. The simulation results indicate that there would be no notable difference in the effect on domestic market protection between BTAs and sectoral exemptions. Both policies are sufficiently effective in maintaining the market share of domestically manufactured steel products in Japan. The differences between the two policies appear in their impacts on other industries and the national economy. Although the implementation of BTAs is less effective than the implementation of sectoral exemptions in mitigating steel production decreases due to emissions trading, it does not lead to serious unfairness in burden sharing among domestic industries or to serious negative economic impacts compared with sectoral exemptions. The negative impacts that sectoral exemptions have on nonexempt sectors and the national economy are emphasized in studies on sectoral exemptions in Germany and Canada (Böhringer and Rutherford, 1998; Rivers, 2010). The implementation of sectoral exemptions under an emissions trading scheme could replace international inequities between regulated countries and countries with less or no regulation with domestic inequities between exempted sectors and nonexempt sectors. Although the implementation of BTAs is still controversial in that it may constitute a disguised restriction on international trade, BTAs are worth considering as an option for ensuring fair international competition via their domestic economic impact.

NOTES

1. For example Kuik and Hofkes (2010) summarize past studies on CGE analysis of carbon leakage and assess the impacts of the EU emissions trading scheme with or without BTAs on carbon leakage. Burniaux, Chateau and Duval (2010) is another recent study that assesses the effects of BTAs implemented by the EU on carbon leakage.
2. According to GATT (1970), BTAs of indirect taxes such as value-added taxes and excise duties are defined as follows:

    Any fiscal measures which put into effect, in whole or in part, the destination principle (i.e. which enable exported products to be relieved of some or all of the tax

3. From the standpoint of an international agreement, Stiglitz (2006) suggests that BTAs are a way to induce non-participating countries to participate in an international climate protection agreement.
4. For more detailed information about the characteristics and methodology of CGE models, see Hosoe *et al.* (2010).
5. The mathematical properties of a CES function are detailed in Silberberg and Suen (2001).
6. www.japanmetaldaily.com/ (accessed 1 July 2010, in Japanese).

REFERENCES

Böhringer, C. and T. F. Rutherford (1998), 'Carbon taxes with exemptions in an open economy: A general equilibrium analysis of the German Tax Initiative', *Journal of Environmental Economics and Management*, **32** (2), 189–203.
Burniaux, J., J. Chateau and R. Duval (2010), 'Is there a case for carbon-based border tax adjustment?: An applied general equilibrium analysis', OECD Economics Department Working Papers, No. 794.
GATT (1970), GATT Document, Report by the Working Party on Border Tax Adjustments, L/3464, adopted on 02.12.1970, published as BISD 18S/97.
Hosoe, N., K. Gasawa and H. Hashimoto (2010), *Textbook of Computable General Equilibrium Modeling: Programming and Simulations*, Hampshire, UK: Macmillan.
International Energy Agency (IEA) (2007), *Tracking Industrial Energy Efficiency and CO2 Emissions*, Paris, France: OECD/IEA.
Kuik, O. and M. Hofkes (2010), 'Border adjustment for European emissions trading: Competitiveness and carbon leakage', *Energy Policy*, **38** (4), 1741–8.
Monjon, S. and P. Quirion (2010), 'How to design a border adjustment for the European Union Emissions Trading System?', *Energy Policy*, **38** (9), 5199–207.
Paltsev, S., J. Reilly, H. Jacoby *et al.* (2005), 'The MIT Emissions Prediction and Policy Analysis (EPPA) Model: Version 4', MIT Joint Program on the Science and Policy of Global Change, Report 125.
Rivers, N. (2010), 'Impacts of climate policy on the competitiveness of Canadian industry: How big and how to mitigate?', *Energy Economics*, **32** (5), 1092–104.
Schumacher, K. and S. Ronald (2007), 'Where are the industrial technologies in energy-economy models? An innovative CGE approach for steel production in Germany', *Energy Economics*, **29** (4), 799–825.
Silberberg, E. and W. Suen (2001), *The Structure of Economics: A Mathematical Analysis*, New York, US: McGraw-Hill.
Stiglitz, J. (2006), 'A new agenda for global warming', *The Economists' Voice*, **3** (7) Article 3, available at www.bepress.com/ev/vol3/iss7/art3/.
Tekko Shinbunsha (2005), *Tekko Nenkan* (Japanese), Tokyo, Japan: Tekko Shinbunsha.
The Japan Iron and Steel Federation (2009), *Handbook for Iron and Steel Statistics* (Japanese), Tokyo, Japan: The Japan Iron and Steel Federation.

15. Effectiveness of the Danish packaging tax

Enian Cela and Shinji Kaneko

I. INTRODUCTION

As in the rest of the world, the EU-15 area is characterized by an increasing use of packaging products expanding everyday to cover new applications. In the period 1998–2006 alone, the amount of packaging placed in the EU-15 market increased by almost 11% (EUROPEN, 2009). At the same time, per capita consumption arouse from 147 to 157 kg on average. In aggregate, the increment is about 10 millions tones. The bulk consists of non-wood packaging, including plastics, paper, paperboard, metal and glass.

Acknowledging the waste problems associated with packaging use, the EU stipulated in 1994 the European Parliament and Council Directive 94/62/EC on packaging and packaging waste. The directive stimulates the member states to take responsive in curtailing packaging demand and use while reserving the right to choose the appropriate policies. At the same time, the directive emphasizes market based policies as more appropriate and their application as more desirable.

Packaging taxation is included in the group of market-based policies. The cases of adoption, however, are scarce. As a matter of fact, the EUROPEN (2000) report identifies only a few countries applying environmental taxation in the case of packaging products. Amongst these examples, Denmark could be considered a 'model'. Not only is the country characterized by a long history of packaging tax application (first taxes being implemented as early as 1978), but coverage is extended to include almost all kinds of packaging products. The purpose in this chapter is not to analyze the entire history of Danish packaging taxation but to focus on the Consolidated Act of 1999 (amended in 2000) which enabled a weight basis product charge taxation system (tax applied per kg of packaging product sold). The charges in this case, addressed the waste problem indirectly by stimulating packaging demand reduction.

It is the purpose of this chapter to investigate the effectiveness of taxation in the case of plastic and paper/paperboard packaging products (charges are different for plastics and paper/paperboard). In this sense, we have chosen two major commodity groups: one incorporating plastic packaging and the other paper/paperboard packaging. The choice of these two groups is not a coincidence. Previous investigations have revealed that environmental tax effectiveness is critically dependable on the demand price elasticity of the targeted commodity (Bailey, 2002). Plastic packaging is known to develop a price-inelastic demand, whilst paper/paperboard is characterized by a price-elastic one. In our approach, we applied two separate trade gravitational regressions: one for each commodity group. In both equations, a tax dummy was included amongst the explanatory variables to capture the impact of the taxation policy. Results reveal that the Danish tax was effective in the case of the paper/paperboard packaging commodity group but proved ineffective in the case of the plastic packaging commodity group.

II. LITERATURE REVIEW

Packaging environmental taxes are applied in two main forms: material levies and product charges (Pearce and Turner, 1993). Material levies are applied on the raw material employed in the production of the packaging product and are designed to improve material efficiency. In this case, the tax purpose is to artificially increase the price of the raw material and induce packaging producers to adopt production technologies that require lower quantities of raw material. Hekkert *et al.* (2000a, 2000b) identify several production technologies belonging to both plastic and paper/paperboard packaging that could be stimulated by applying such taxes. Product charges, on the other hand, are applied on the final product and the purpose is to reduce final demand. According to Pearce and Turner (1993), product charges could induce a change in the purchased quantity of virgin packaging by either stimulating a source reduction demand and/or increasing the recycling rate. In our analysis, the Danish packaging tax object of investigation is in fact a product charge. The packaging user pays the tax at the moment of purchase.

Before investigating the tax-effectiveness in each case, it is necessary to point out certain favorite circumstances that might work in favor of achieving successful implementation. First, since we are dealing with industrial packaging (purchased by firms to package the product they sell in the market), effectiveness would require for the tax not to be pushed back to the final consumer in the form of increased price for the packed products. A study commissioned by the European Commission and conducted by ECOTEC Research and Consulting (ECOTEC, 2001) on the *Economic and environmental implications of the use of environmental taxes and charges in the European Union and its member states* investigates such eventuality in the case of the Danish product charge. The study reports that there is no evidence pointing toward a tax push-back to the final consumer.

Another very crucial issue that could seriously affect the tax-effectiveness or ineffectiveness is related to the demand price elasticity of the commodity. A price-elastic demand would increase the probability of successful product charge implementation, whilst a price-inelastic demand would reduce such probability. In the case of paper/paperboard packaging, evidence suggests the presence of a highly price-elastic demand. Suohen (1984) analyzes the demand for three categories of paper and paperboard in the case of European Economic Community countries using historical data from 1968 until 1980. Results reveal demand related negatively to the price in the case of the category denoted as 'Other Paper and Paperboard', which includes packaging. Prestemon and Buongiorno (1993) replicate with a sample of 24 OECD countries analyzing the demand for three paper and paperboard categories (one of them includes packaging). Once again, the results showed the presence of negative correlation between demand and price in the case of the packaging category. Chas-Amil and Buongiorno (2000) applied a country-by-country analysis for 14 members of the EU using short-erm elasticities of demand for forest products. In the case of Denmark, results revealed a negative correlation between demand and price with statistical significance at 1% levels of confidence. There are also examples of global demand modeling. Simangunson and Buongiorno (2001), using an international demand equation for forest products and applying four different methodologies, define the price elasticities in the case of nine paper- and paperboard-related products (including paper/paperboard packaging). According to each method, the demand for paper/paperboard packaging is price elastic and statistically significant.

Regarding plastic packaging, there is a shortage of studies incorporating demand modeling. However, the few studies available suggest that demand is characterized by price inelasticity. Palmer, Sigman and Walls (1996) in their study on the US market of packaging products found a price-inelastic demand. Furthermore, Zhang and Buongiorno (1998) define plastic packaging as a luxury commodity featuring a demand not affected by price variations. At the same time, the EuPC (European

Plastic Converters, 2010), based on plastic packaging market observations, determine that despite the continuous price pressures that have characterized the European market, demand for plastic packaging continues to grow relentlessly and unaffected.

Considering what was said so far, we expect the Danish tax policy to be effective in the case of paper/paperboard packaging but not in the case of plastic packaging. In the next section, we explain the methodology, the model and data employed in our analysis.

III. CONCEPT, MODEL AND DATA

As mentioned earlier, our analysis focuses on two major commodity groups:

- boxes, cases, crates and similar articles of plastic hereon referred as the plastic group;
- paper and paperboard for packaging purposes hereon referred as the paper/paperboard group.

In both cases we are dealing with major packaging commodity groups purchased and used by various industrial sectors. According to Statistics Denmark (2010), in both cases the major purchaser is the domestic food, beverage and tobacco industrial sector, accounting for more than 65% of total purchases by weight. In comparison, the shares of the other sectors are marginal.

As the title also suggests, our analysis is constrained to the imports of each commodity group. The reason is that according to the data (Statistics, Denmark 2010), Denmark is a major importer and not a producer of both commodity groups. However, the partner countries' sets are different in each case. We chose the trade gravitational model approach considering its popularity and relative simplicity in analyzing trade impacts. Two different models are applied, one for each commodity group. In the case of the plastic group, the equation is developed as follows:

$$\ln IMP_{jt} = \beta_1 \ln SFBTt + \beta_2 \ln RGDPj_t + \beta_3 \ln RPRI_{jt}$$
$$+ \beta_4 DIST_j + \beta_5 TI_{jt} + \beta_6 BOR + \beta_7 TAX + \epsilon \qquad (1)$$

The dependent variable represents quantity imports (expressed in kg) of Denmark from partner country j in period (year) t. As the tax policy is supposed to decrease virgin packaging demand, decreased quantity imports would be the signal of effective policy implementation. The first explanatory variable (ln *SFBT*) is defined as sales from the Danish food, beverage and tobacco industrial sector and is defined as a proxy of demand. As the sector stands as a major purchaser of plastic packaging, we expect it to be the most affected by the policy (should the policy have any effect). Sales are expressed in real 2005 Danish Kroners (DKr) and are seasonally adjusted (Statistics Denmark, 2010). The adjustment enables to correct for any possible market price related, inflationary or seasonal effect. We expect a positive sign and statistical significance for the coefficient of this variable.

RGDP is defined as Real GDP of partner country *j* in period *t* and represents a supply proxy. We assumed that the larger the partner country, the higher would be its producing and exporting potential of plastic packaging and, therefore the more Denmark would import from it. The variable is expressed in real 2000 US$ once again adjusting for price effects. *RPRIjt* is defined as real annual average price of commodity imports in year *t* expressed in real 2000 DKr per kg of product. This is the price domestic purchasers have to pay for each kg of imported plastic commodity they buy. The inclusion of the variable in this case is made for two main purposes. The first is to obtain a confirmation on the nature of price elasticity of demand for the commodity group in question. The second is to separate the price effect from the tax effect and judge whether import changes are due to pure market related price variations and/or to the adoption of the tax policy. However, judging from the literature, we expect the coefficient to be statistically insignificant. The variable ln *DIST* reflects the distance in km between the Danish capital Copenhagen and the partner country *j* capital city. This is a proxy of transportation costs which could have certainly affected trade flows. The next two explanatory variables are both dummies. *TI* is denoted as trade integration proxy taking values 1 for the periods when the partner country *j* was a member of the European Union. It is commonly acknowledged that EU represents a very well-integrated market where there is ample freedom in the movement of goods, capital and labor. Therefore, member states are bound to conduct a higher intensity of trade with one another compared with the rest of the world. *BOR* is a dummy variable identifying common borders between Denmark and the partner country *j* as a proxy of regional trade. For both variables, we expect a positive sign and statistically significant coefficients. The last variable is a dummy for the taxation policy. The dummy takes values 1 from 2001 and on the years when the packaging tax was applied in its entire form. We mentioned earlier that the tax policy was first proposed in 1999. However, several amendments were proposed and adopted during 2000. For this purpose, we assume 2001 as the first year of full implementation. In the analysis we have included 19 partner countries chosen as the major suppliers of the plastic packaging commodity group in question. The period of analysis stretches from 1994 until 2007. The choice is not random. First, we made sure to include a sufficient number of years from both before and after the policy implementation in order to better capture the taxation effect. Secondly, we managed to leave out time periods featuring major economical shocks that could have certainly affected trade relationships between countries. We can mention here the dissolution of the Soviet block in early 1990s and the economical and financial crisis which has been pounding Europe and the whole world since 2008.

In the case of paper/paperboard, the equation is similar but with just one difference. We did not include the RGDP of partner countries as supply proxy. The equation is as follows:

$$\ln IMP_{jt} = \beta_1 \ln SFBTt + \beta_2 \ln RPRI_t + \beta_3 \ln DIST_j$$
$$+ \beta_4 TI_{jt} + \beta_5 BOR_j + \beta_6 TAX_t + \epsilon \qquad (2)$$

The other variables are the same ones applied in the plastic group equation. The difference is that here we expect the real price and tax variable coefficients to produce negative signs and statistical significance. In this case, the sample includes 13 major partner countries. The analysis is conducted for the same period of time: 1994–2007. Data on imports and sales from the food, beverage and tobacco sector were obtained from the Danish Statistics database. Calculations for real prices were conducted based on data from the same source. Real GDP data for partner countries in the case of the first equation were obtained from the International Monetary Fund (IMF) database.

IV. PRELIMINARY TESTS

Before conducting regression analysis, it is crucial to establish whether the dependent and independent variables are stationary. Non-stationary variables could produce spurious and therefore unreliable results. Unit root analysis was conducted using the Augmented Dickey-Fuller and Phillips-Perron methods. Results for Equation 1 variables are shown in Table 15.1; results for Equation 2 variables are shown in Table 15.2.

Table 15.1 Unit root test results for Equation 1 variables (at level and first difference)

Unit root	Method	lnIMP	lnSFBT	lnRGDP	lnRPRI
Level	ADF	75.5447 (0.0003)	0.7440 (1.0000)	4.8478 (1.0000)	81.0388 (0.0001)
	PP	68.1463 (0.0019)	0.0423 (1.0000)	3.5848 (1.0000)	80.5924 (0.0001)
First difference	ADF		81.4951 (0.0000)	54.3133 (0.0419)	
	PP		84.2238 (0.0000)	56.6539 (0.0263)	

The tests are conducted at individual intercept using Schwarz automatic lag selection, Bartlett method and Newey-West Automatic bandwidth selection. Probabilities are displayed in parentheses.

Table 15.2 Unit root test results for Equation 2 variables (at level and first difference)

Unit Root	Method	lnIMP	lnRPRI
Level	ADF	48.6479 (0.0045)	33.0876 (0.1596)
	PP	53.3382 (0.0012)	14.7022 (0.9625)
First difference	ADF		53.0909 (0.0013)
	PP		54.5245 (0.0009)

The tests are conducted at individual intercept using Schwarz automatic lag selection, Bartlett method and Newey-West Automatic bandwidth selection. Probabilities are displayed in parentheses.

The ln *SFBT* variable appears in both equations. For this reason, the unit root test results for this variable are shown only in Table 15.1. It can be observed that the dependent variable is stationary at level in both cases. However, there are explanatory variables non-stationary at level. In the case of the first equation, there are two of them: ln *SFBT* and ln *RGDP*. In the case of the second equation, once again there are two non-stationary explanatory variables: ln *SFBT* and ln *RPRI*. In this case, Unit Root test is conducted at first difference for the explanatory variables that are non-stationary at level.

The results provide a most singular case. In both equations, the dependent variable is stationary at level. Two of the explanatory variables are stationary at first difference. Therefore, it is not possible to apply co-integration technique as that would require both dependent and independent variables to be integrated in the same order (Engle and Granger, 1987). Pagans and Wickens (1989) argue that in the case of a dependent variable being stationary at level, there must be at least two independent variables of the same order of integration for the equation to be correctly specified. Furthermore, they suggest a way to determine whether the model retains specification. This is achieved by running a unit root test of the residual. If results show that the residual is stationary at level, the model retains the correct specification; otherwise is misspecified. After running the models and obtaining the results, it is necessary to run unit root test for the residuals in each case.

V. RUNNING THE MODELS

Having conducted the preliminary stationarity tests, we run the models first applying ordinary least squares. The results are shown in Table 15.3.

Table 15.3 Ordinary least squares regression results

Variable	Equation 1	Equation 2
ln *SFBT*	0.824411 (5.404678)***	1.234554 (5.043152)***
ln *RGDP*	0.328075 (6.515719)***	
ln *RPRI*	0.496960 (0.593624)	-0.994493 (-1.290145)
ln *DIST*	-0.838804 (-6.095214)***	-2.077689 (-4.726748)***
TI	-0.540391 (-2.184699)***	-1.742968 (-1.927783)*

Variable	Equation 1	Equation 2
BOR	1.434601 (6.109218)***	0.883209 (1.695456)*
TAX	0.001623 (0.9924)	-0.626747 (-1.694679)*
R^2	0.479352	0.341394
Adjusted R^2	0.467290	0.321793
DW stat	0.467353	0.489420

t-stat shown in parenthesis. * significance at 10% confidence intervals; *** significance at 1% confidence intervals.

The next step would be to conduct heteroskedasticity and autocorrelation tests for both models. Applying the White's General Heteroskedasticity Test, we obtained an nR^2 value of 36.15738 in the case of Equation 1. The 5% critical χ^2 distributed value is 14.0671. Since the obtained nR^2 value is higher than the critical χ^2 distributed value, we can reject the null hypothesis on no heteroskedasticity in the case of Equation 1. In Equation 2, the obtained nR^2 value is 20.88, once again higher than the 5% critical χ^2 distributed value of 12.59158. Once again the null hypothesis of no heteroskedasticity is rejected. Apart from the problem of heteroskedascity, in both equations is also encountered the problem of serial autocorrelation. From Table 15.3 we can observe that the Durbin-Watson statistics are 0.467353 and 0.489420 respectively, announcing the existence of positive autocorrelation. Therefore, it is necessary to correct for both heteroskedasticity and autocorrelation using Seemingly Unrelated Regressions (SUR). However, this method presents an important limitation. The standard errors are downward biased (Beck and Katz, 1995) and is suggested the application of bootstrapped standard errors (Messemer and Parks, 2004). On the other side, Atkinson and Wilson (1992) argue that the standard errors downward bias is present also in the bootstrapped regression. Keeping that under consideration, both SUR and bootstrapped methods cannot dominate one-another being both equally valid. Results using the SUR method are shown in Table 15.4.

Table 15.4 SUR analysis results

Variable	Equation 1	Equation 2
ln *SFBT*	0.77 (9.53)***	1.30 (23.69)***
ln *RGDP*	0.85 (6.51)***	
ln *RPRI*	−0.32 (−1.33)	−0.93 (−5.02)***
ln *DIST*	0.58 (−3.98)***	−2.18 (−19.55)***
TI	−0.33 (−5.95)***	−2.15 (−19.55)***
BOR	1.26 (4.47)***	0.77 (7.37)***
TAX	0.07 (1.01)	0.54 (−5.54)***
R^2	0.19	0.95
Adjusted R^2	0.17	0.95
DW stat	1.97	1.62

t-stat shown in parenthesis. *** significance at 1% confidence intervals.

We have therefore reached a situation where heteroskedasticity and auto-correlation problems have been corrected. The next step is to determine model specification by running unit root test of the residuals with the same methods applied beforehand for the variables. Tests indicate residuals being stationary at level retaining model specification in both equations. In the end, we also performed Variance Inflation Factors (VIF) (Fox and Mon-ette, 1992) to test for the eventuality of multicollinearity. Results rejected such eventuality in both models. Having performed all necessary tests, in the next section we explain the obtained results.

VI. RESULT EXPLANATION

Results from Table 15.4 reveal that the lnSFBT variable produced the expected sign and statistical significance. Sales from the food, beverage and tobacco sector are positively correlated with the imports belonging to both packaging groups. The magnitude, however, is higher in the case of paper/

paperboard imports. In the case of Equation 1, the lnRGDP variable proxy of supply also produced the expected sign and statistical significance. The distance variable as proxy of transportation costs also produced the expected sign and statistical significance in both models. Despite the static nature of the variable, results revealed statistical significance with a magnitude higher in the case of paper/paperboard packaging. The common border dummy is also significant and with the expected sign revealing the importance of regional trade in both models. The Trade Integration dummy, on the other hand, did not produce the expected sign but it produced the expected statistical significance.

But let us focus on the two variables of the outmost importance in our investigation: price and the tax dummy. The results confirmed what was expectable from the literature findings. In the case of plastic packaging, the price-inelastic demand is confirmed by the statistical insignificance of the real price variable in Equation 1. A condition of demand being non-sensitive to price changes occurs. In the case of paper/paperboard packaging, the opposite is reflected. Import demand is very sensitive to price alterations. That is confirmed by the magnitude and statistical significance of the coefficient.

The other variable of paramount importance in our case is the tax policy dummy variable. Once again, results show different impact in each case. In the case of the plastic packaging commodity group, the product charge seems to have not affected the import quantity. The statistical insignificance of the coefficient stands as proof of this fact. In the case of paper/paperboard packaging the result is entirely different. The high statistical significance reveals that the tax policy managed to affect imports in the intended manner. Furthermore, the effect is separated from that of price changes proving that in the end, it was a good idea to include both variables in the equations. By doing so, we showed that paper/paperboard imports are affected by both price and tax adding to the model reliability.

VII. CONCLUSIONS AND DISCUSSIONS

This chapter represents an attempt to investigate the effectiveness of the Danish packaging product charge policy on quantity imports belonging to two different packaging commodity groups: one composed of plastic packaging and the other of paper/paperboard packaging. The commodity groups are different in the sense that the plastic group is characterized by a price-inelastic demand whereas paper/paperboard packaging features a price-elastic demand (in our case, import demand). This represents a crucial difference that we expected to have an important impact in stimulating tax policy effectiveness.

The approach included the application of two separate trade gravitational regression models: one equation for the plastic group and another for the paper/paperboard group. In both cases, imports were defined as the dependent variable whilst a tax policy dummy was included amongst the explanatory variables. At the same time, explanatory variables encompassed demand, supply, price, transportation cost and regional trade proxies. The real price variable was included with the premeditated intent to separate its effect from that of the tax.

Regarding the effectiveness of the taxation policy, the results are mixed. In fact, success is experienced in the case of paper/paperboard imports where the charge managed to induce import reduction. In the case of plastic packaging, the tax policy did not produce any effect on import demand. These results confirm once again the vulnerability of environmental taxation to the demand price-elasticity of the targeted commodity. In the case of plastic packaging, the existence of price-inelastic demand stands as a barrier in the way of successful tax implementation. Therefore, it would seem that product charges are not the appropriate response, at least in the case of the plastic packaging commodity group.

Under these circumstances, two alternative policy implications could be advanced. The first one, as suggested from the literature, would be the adoption of a material levy system that improves efficiency. However, in the case of a major importer country that is not engaged in the production, it would be pointless advancing such policy. The second option is related to price inelasticity reversal. In other words, demand for new plastic packaging has to be turned from price-inelastic to price-elastic. A primary way to achieve this is through recycling. As a matter of fact, in the case of paper/paperboard packaging, the ability to increment recycling rates might have increased recycled packaging use and reduced the demand for virgin packaging (see EUROPEN, 2009). In the case of plastic packaging, due to low recycling rates, demand price inelasticity does not work in favor of demand reductions through taxation. Therefore, economical (and taxation) policies have to be synchronized with scientific research that aims at increasing the recycling rate of plastic packaging.

REFERENCES

Atkison S.E. and P.W. Wilson (1992), 'The bias of bootstrapped versus conventional standard errors in the general linear and SUR models', *Economic Theory*, **8**, 258–75.

Bailey I. (2002), 'European environmental taxes and charges: economic theory and policy practice', *Applied Geography*, **22**, 235–51.

Beck N. and J.N. Katz (1995),'What to do (and not to do) with time-series cross-section data', *The American Political Science Review*, **89** (3), 634–47.

Chas-Amil M.L. and J. Buongiorno (2000), 'The demand for paper and paperboard: econometric models for the European Union', *Applied Economics*, **32** (8), 987–99.

ECOTEC (2001), *Study on the economic and environmental implications of the use of environmental taxes and charges in the European Union and its member states*, Brussels: ECOTEC.

Engle R.F. and C.W.J. Granger (1987), 'Co-integration and error correction: representation, estimation and testing', *Econometrica*, **55** (2), 251–76.

EUROPEN (2000), *Economic instruments in packaging and packaging waste policy*, Brussels: EUROPEN.

EUROPEN (2009), *Packaging and packaging waste statistics 1998–2006*, Brussels: EUROPEN.

Fox J. and G. Monette (1992), 'Generalized collinearity diagnostics', *Journal of the American Statistical Association*, **87** (417), 178–83.

Hekkert M.P., L.A.J. Joosten and E. Worrell (2000a), 'Reduction of CO_2 emissions by improved management of material and product use: the case of primary packaging', *Resources, Conservation and Recycling*, **29**, 33–64.

Hekkert M.P., L.A.J. Joosten and E. Worrell (2000b), 'Reduction of CO_2 emissions by improved management of material and product use: the case of transport packaging', *Resources, Conservation and Recycling*, **30**, 1–27.

Messemer C. and R.W. Parks (2004), 'Bootstrap methods for inference in a SUR model with autocorrelated disturbances', *University of Washington Economics Working Paper No. UWEC-2004–24*.

Pagans A.R. and M.R. Wickens (1989), 'A survey of some recent econometric methods', *The Economic Journal*, **99** (398), 962–1025.

Palmer K., H. Sigman and M. Walls (1996), 'The cost of reducing municipal solid Waste', *RFF Discussion Paper*, 96–35.

Pearce D.W. and R.K. Turner (1993), 'Market-based approaches to solid waste management', *Resources, Conservation and Recycling*, **8**, 63–90.

Prestemon J. and J. Buongiorno (1993), *Elasticities of demand for forest products based on time-series and cross-sectional data*, Groupe de Reserche en Economie des Produits Forestiers, University of Bordeaux, Seminar 24–25 June 1993.

Simangunsong B.C.H. and J. Buongiorno (2001), 'International demand equation for forest products: a comparison of methods', *Scandinavian Journal of Forest Research*, **16** (2), 155–72.

Suhonen T. (1984), *Price variable in dynamic consumption, models of selected paper products: a pooled cross-section and time series analysis*, Helsinki: Jaakko Pöyry International Oy.

Zhang Y. and J. Buongiorno (1998), 'Paper or plastic? The United States' demand for paper and paperboard in packaging', *Scandinavian Journal of Forest Research*, **13** (1), 54–65.

WEB REFERENCES

Statistics Denmark, www.statbank.dk, accessed 5 January 2010.

Danish Ministry of Taxation, 'Consolidated Act on taxes on certain types of packaging, bags, disposable tableware and PVC foils', www.skm.dk/foreign/english/2087.html, accessed 10 February 2010.

European Plastic Converters, 'Plastics packaging markets in Europe', www.plasticsconverters.eu/markets/packaging, accessed 21 January 2010.

International Monetary Fund, 'World economic outlook databases', www.imf.org/external/ns/cs.aspx?id=28, accessed 18 December 2009.

Index

3 5282 00720 2842